International Workshop on Water-Jet Propulsion (IWWP2022)

AA005707

Online
30 June – July 2022

ISBN: 978-1-7138-8310-4

Printed from e-media with permission by:

Curran Associates, Inc.
57 Morehouse Lane
Red Hook, NY 12571

Some format issues inherent in the e-media version may also appear in this print version.

Printed with permission by Curran Associates, Inc. (2024)

For permission requests, please contact The Royal Institution of Naval Architects at the address below.

The Royal Institution of Naval Architects
8-9 Northumberland Street
London, WC2N 5DA
United Kingdom

Phone: 020 7235 4622
Fax: 020 7259 5912

publications@rina.org.uk

Additional copies of this publication are available from:

Curran Associates, Inc.
57 Morehouse Lane
Red Hook, NY 12571 USA
Phone: 845-758-0400
Fax: 845-758-2633
Email: curran@proceedings.com
Web: www.proceedings.com

160 YEARS OF HISTORY, KNOWLEDGE AND EXPERIENCE

THE ROYAL INSTITUTION
OF NAVAL ARCHITECTS

International Workshop on Water-Jet Propulsion (IWWP2022)

30th June - 1st July

Online

In association with:

MARIC

THE ROYAL INSTITUTION
OF NAVAL ARCHITECTS

INTERNATIONAL WORKSHOP

Water-Jet Propulsion (IWWP2022)

30th June - 1st July 2022, Online

THE ROYAL INSTITUTION OF NAVAL ARCHITECTS
8-9 Northumberland Street
London WC2N 5DA

Telephone: 020 7235 4622
Fax: 020 7259 5912

CONTENTS

NUMERICAL SIMULATION OF FLOW CHARACTERISTICS OF WATER-JET ASSEMBLY UNDER REVERSE STATE

Y Long, J Zhong and **R Zhu** National Research Center of Pumps of Jiangsu University, China
J CHEN and **Y Zhang,** Marine Design and Research Institute of China, China

SUMMARY

Although the water-jet assembly has a lot of shipping applications after its development and has a good reputation, it is not known whether this creation 40 years ago can withstand the verification of modern CFD technology. Through the CFD calculation of water-jet assembly under forward state and reverse state, the internal flow of water-jet assembly during forward state and reverse state is studied. The results show that the flow characteristics of the water-jet assembly in the reverse state are different from the original design intention. There are reverse jets on both sides of the gap between the nozzle and the built-up rudder, and the non-twisted blades of the guide vane cannot completely eliminate the circumferential velocity of the fluid after the work of the impeller. Therefore, the built-up rudder and guide vane of water-jet assembly may need to be further optimized.

NOMENCLATURE

D	Nominal impeller diameter (mm)
D_2	Nozzle diameter (mm)
E_{nD}	Energy coefficient (dimensionless number)
g	Gravitational acceleration (m/s^2)
H	Head (m)
n	Rotation speed (rev/min)
η	Efficiency

1. INTRODUCTION

Unlike propeller propulsion, which was widely used in the 19th century, although the concept of water-jet propulsion was proposed earlier, it is a special propulsion method that has only been developed and applied in the past 40 years. It uses the reaction force of the water-jet from the propulsion pump to push the ship forward [1]. There are two main forms of water-jet propulsion: Inboard water-jet propulsion and outboard water-jet propulsion. With the development of high-performance ships, water-jet propulsion has been more and more widely used because of its good mobility and manoeuvrability, shallow draft, low vibration and noise, high propulsion efficiency at high speed and strong anti-cavitation ability [2].
The externally suspended water-jet assembly studied in this paper is an energy-saving propulsion device originally created by MARIC. Its shape is similar to the duct propeller. Most of it is installed on engineering ships with tunnel tail line. The propulsion pump and built-up rudder are the main body of the device, as shown in Figure 1.

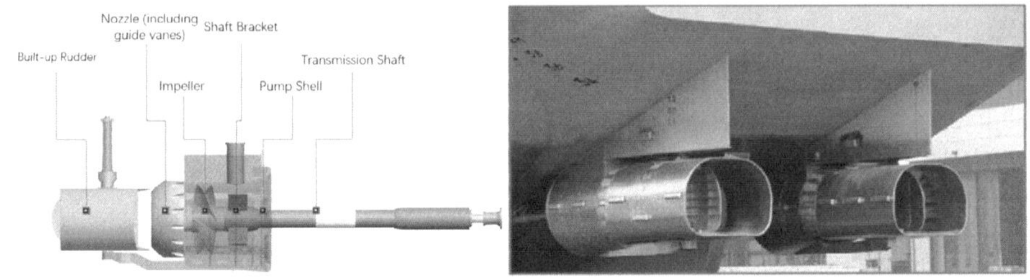

Figure 1. Water-jet assembly (Left) and physical objects on engineering ships (Right)

The efficiency of water-jet propulsion is closely related to the operating efficiency of the pump, so many experts and scholars have conducted a lot of in-depth research here. The operating efficiency of the pump is mainly related to the impeller and guide vane. Chang Shuping et al [3] studied the influence of leading edge and trailing edge design of the impeller on the performance and flow characteristics of water-jet propulsion pump. Zhang Yan et al [4] studied and analysed the effects of different geometric parameters on the hydraulic performance of water-jet propulsion pump, such as the number of shaft bracket, the placement angle of shaft bracket blade outlet, the dynamic and static spacing between shaft bracket and impeller, the tip clearance of impeller. Han Wei et al [5] numerically simulated the water-jet propulsion pump models with different axial clearance coefficients between guide vanes and impellers, and studied the changes of pump external characteristics, internal flow energy conversion and thrust characteristics under different axial clearance coefficients. Gong Jie et al [6] studied the unsteady hydrodynamic performance of axial-flow water-jet impeller in a certain speed range by numerical simulation. Lu Yiming et al [7] studied the influence of impeller tip clearance on the working characteristics of high-speed water-jet propeller by numerical simulation. In addition, he also studied the working characteristics under different cruise speeds. Many other scholars have conducted targeted research on the tip leakage vortex of the impeller [8-10], revealing the transient characteristics and evolution process of the tip leakage vortex.

The propulsive efficiency of water-jet propeller is not only related to pump efficiency, but also related to system efficiency. The system efficiency is mainly affected by the inlet velocity ratio and pipeline loss coefficient. The nozzle area determines the inlet velocity ratio, and the pipe loss is composed of nozzle loss and inlet channel loss. Shi Jun and Zhang Wen et al [11,12] established a variety of channel models with different inlet area and inlet length. Through the numerical calculation of the models under different inlet velocity ratio (IVR), the flow characteristics of the flow field in the channel are obtained, the effects of different inlet area and inlet length on the flow performance of water jet propulsion pump inlet channel are analysed from the aspects of outlet velocity weighted average angle, wall cavitation and channel overflow loss. Proper stagnation point location and good outlet uniformity are of great benefit to improve the performance of water-jet pump. Ji Guorui et al [13] studied the influence of lip parameters of water-jet propulsion inlet channel on outlet non-uniformity and stagnation point location by CFD method, which provides guiding value for the selection of lip size.

Under the in-depth research of many scholars, the efficient and stable operation of water-jet propulsion pump has been basically guaranteed. However, the above studies are almost all aimed at the water-jet propulsion pump. There is less research on the steering reversing mechanism of water-jet propeller, and there is less research on the built-up rudder of water-jet assembly which is the object of this paper. Due to the huge capital investment required for the model test of steering reversing mechanism, the experimental research on steering reversing mechanism in China is seriously insufficient at present. Some people conducted numerical simulation on two conventional types of steering reversing mechanism and verified the effectiveness of the numerical simulation method combined with the published experimental data [13-15]. Liu Xueqin [16] has studied and analysed the force and structural strength of the built-up rudder of the water-jet assembly, but she seems to only verify the rationality of the combination in structural strength but has not conducted a more in-depth study on the flow structure optimization of the water-jet assembly. Therefore, this paper hopes to carry out further research and analysis on the internal flow field of the assembly, so as to improve the flow structure of the assembly, improve the propulsion efficiency of water-jet assembly.

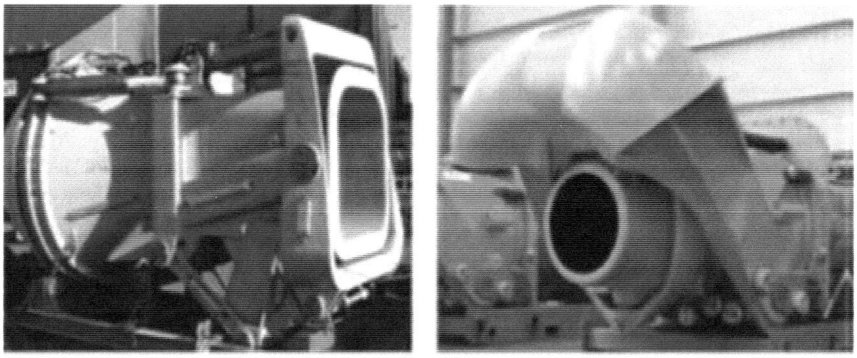

Figure 2. Two conventional types of steering and reversing mechanisms: integrated type(Left); Split type(Right)

2. STRUCTURAL COMPOSITION AND CALCULATION SETUP

2.1 PROPULSION PUMP

The propulsion pump is mainly composed of pump shell, shaft bracket, impeller, nozzle (including guide vane) and tail shaft system. The basic parameters of the model are shown in Table 1. The propulsion pump is suspended at the stern through the connection between the pump shell and the stern.

Table 1. Model geometric parameters

parameter	Number of shaft bracket blades	Number of impeller blades	Number of guide vane blades	Nozzle diameter(D_2)
value	3	4	5	600mm

2.2 BUILT-UP RUDDER

As shown in Figure 3, the built-up rudder is composed of rudder shell, split-flow plate, and middle rudder plate. The middle rudder plate and rudder shell can rotate respectively to form different flow channels and control the direction and flow distribution of jet. The rudder shell is a hollow box welded structure with an oval cross section. When the middle rudder plate rotates to a certain angle, it can fit with the rudder shell and side plate to close the flow channel [1,2]

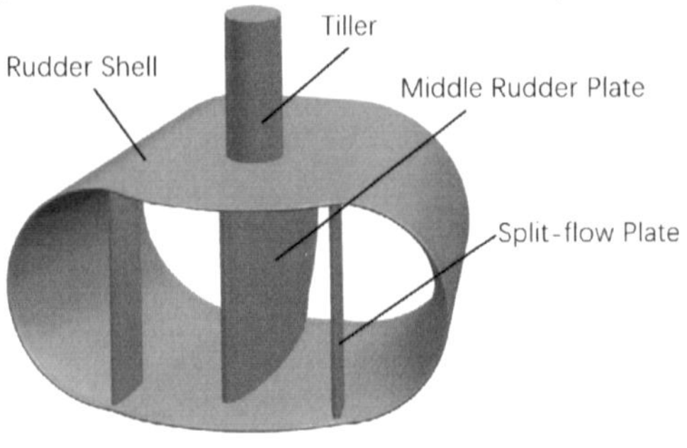

Figure 3. Three-dimensional structure of built-up rudder

2.3 BUILT-UP RUDDER CONFIGURATION DURING FORWARD AND REVERSE STATE

As shown in Figure 4, the rudder shell and middle rudder plate do not rotate when the ship is in the forward state. When sailing backwards, the middle rudder plate deflects to the outer chord until it is closed with the rudder shell to form a closed cavity, it can reverse the jet and steer the ship backwards.

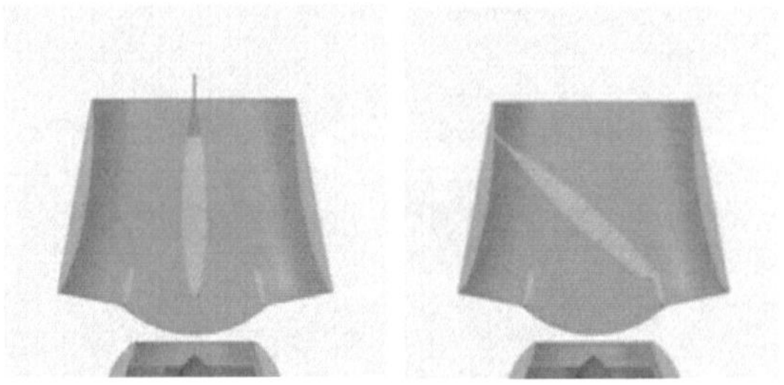

Figure 4. Built-up rudder configuration under different navigation states: Forward state (Left); Reserve state (Right)

2.4 NUMERICAL CALCULATION METHOD

According to the paper [17], the geometric model of the stationary region of the built-up rudder is established. After the actual calculation, it is found that reducing the size of the stationary region of the built-up rudder will not have a visible significant impact on the flow field in the rudder. Therefore, in order to reduce the number of grids and reduce the calculation consumption, a cylindrical fluid region with a diameter of 5D and a length of 10d is finally established to capture the wake flow field behind the rudder, as shown in Figure 5.

The hybrid grid scheme is adopted for the grid division before calculation. Turbogrid is used for the hexahedral grid generation of the shaft bracket, impeller, and guide vane. ICEM is used for the hexahedral grid generation. The static area of the built-up rudder adopts the unstructured grids. The specific grid is shown in Figure 6.

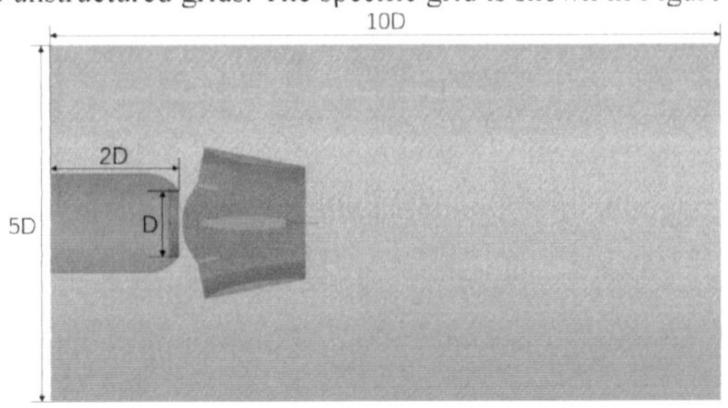

Figure 5. Calculation domain of built-up rudder

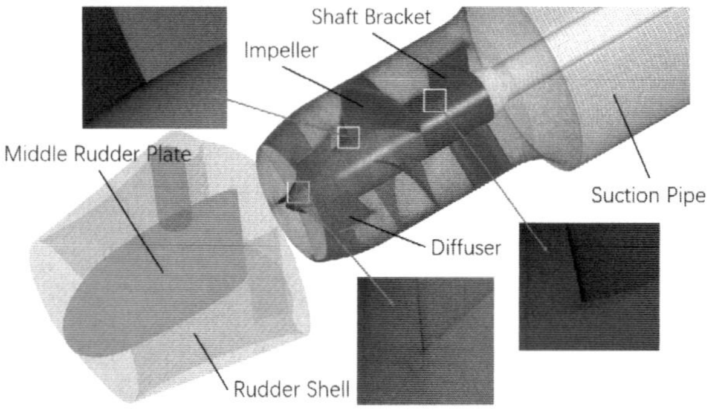

Figure 6. Fluid domain grid of water-jet assembly

The ANSYS CFX 19.0 is used to complete the flow calculation of water-jet assembly. In the calculation, the liquid phase is 25°C water, its density is 997 kg/m^3, and the dynamic viscosity is 8.899×10^{-4}kg/(m·s), using k-ε Turbulence model. In this paper, only the designed operating condition of water-jet propulsion pump is calculated, the mass flow rate is used as the inlet boundary condition, and the boundary of the calculation domain of the built-up rudder is set as opening [18]. The wall boundary condition adopts non slip wall. The calculation domain of the impeller is set to rotating, the speed is 361 rev/min, the blade and hub are set to rotating, the wall velocity of the shroud is set to counter rotating wall. The guide vane, shaft bracket, suction pipe and built-up rudder are set to stationary. The interface between rotating parts and stationary parts is set to frozen rotor. Advection scheme is set to high resolution, and the preset number of iterations is 1000.

The head H data obtained by numerical calculation is transformed dimensionless according to the following formula according to the provisions of the International Standards Committee (IEC):

$$E_{nD} = \frac{gH}{n^2 D^2} \tag{1}$$

Where energy coefficient E_{nD} is a dimensionless number. H, n, D and g represent head, rotation speed, nominal impeller diameter and gravitational acceleration.

Adjust the grid density to generate 5 sets of grid schemes for grid independence verification. The scheme details and calculation results are shown in Figure 7. As can be seen from the figure, with the continuous densification of the grid, the changes of energy coefficient E_{nD} and efficiency η is small. In order to capture the finer flow field structure as much as possible and comprehensively consider the calculation consumption, scheme 4 is finally selected for calculation, with a total grids of 11.29 million.

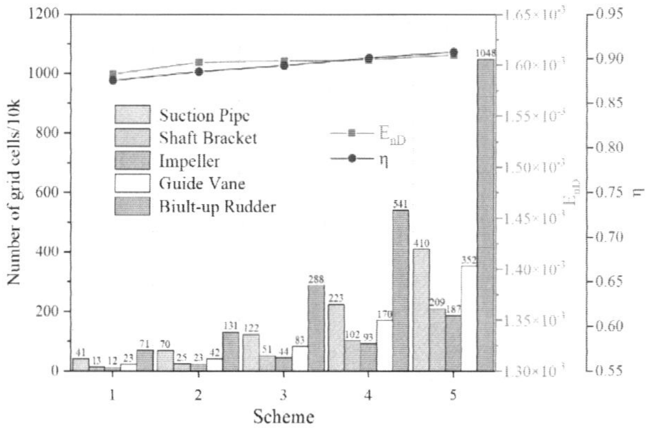

Figure 7. Grid schemes and independence test

3. RESULT ANALYSIS

In the current research, the flow characteristics of the combination of forward state and reserve state are extracted and analysed.

3.1 INTERNAL AND EXTERNAL FLOW OF BUILT-UP RUDDER DURING FORWARD AND REVERSE STATE

Figure. 8 is the streamline, velocity distribution and pressure distribution in the axial plane perpendicular to the rotating axis of the built-up rudder during forward state and reverse state. After the impeller working, the fluid pressure increases, and then through the contraction nozzle, the speed increases, and the pressure decreases. During forward state, the jet from the nozzle is ejected backward through the rudder shell, and there is jet entrainment in the gap between the propulsion pump and the rudder. During reverse state, the middle rudder plate and rudder shell form a closed cavity. The pressure rise is caused by the blockage of the flow channel. The pressure rise area extends from the closed cavity to the impeller, and the jet at the nozzle is sprayed reversely from the gap to realize the reverse sailing of the ship. According to the original design intention, the reverse jet should be ejected only to one side. However, when the propulsion pump operates at the designed operating condition, the jet is ejected on both sides of the built-up rudder, and there is reflected flow in the front and rear of the split-flow plate in deflecting side. The reason may be that the distance between the nozzle and the built-up rudder is too large or there is a problem in the design of the split-flow plate.

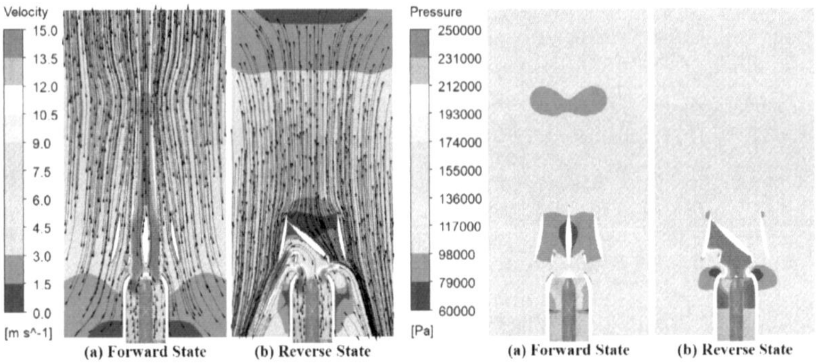

Figure 8. Streamline, velocity distribution and pressure distribution of built-up rudder flow field

3.2 INTERNAL FLOW OF GUIDE VANE DURING FORWARD AND REVERSE STATE

Different from the conventional built-in water-jet propulsion, the reflected water flow of the water-jet propulsion assembly during reverse state is ejected from the gap between the nozzle and the built-up rudder. Because the position of the reflected water flow is close to the guide vane of the propulsion pump, it may have some impact on the internal flow of the guide vane.

Figure. 9 is the position of each guide vane in the water-jet assembly, and Figure. 10, Figure. 11 and Figure. 12 are the pressure distribution, velocity distribution and streamline distribution in the guide vane respectively. It can be seen from the figure that the overall pressure in the guide vane increases while the velocity distribution is not much different during reverse state, but in the area close to the hub, due to the increase of outlet pressure, the reverse pressure gradient near the suction surface of the guide vane increases and the velocity decreases, and finally the reflux vortex appears at the turn of the arc surface of the suction surface. In addition, it can be seen from the streamline diagram that the inflow angle of the fluid after the impeller working is not consistent at different spans of the guide vane, and the angle near the hub is larger. However, since the guide vane of the water-jet propulsion pump assembly is a non-twisted blade, the angles on different spans of the blade are consistent. The final result is that the fluid near the hub is easy to produce flow separation due to the large angle of attack, resulting in backflow vortex on the suction surface of the blade, the fluid near shroud still has a certain circumferential velocity after flowing out of the guide vane, as shown in Figure. 13, and part of the rotational kinetic energy cannot be utilized. Therefore, there may be some problems in the design of guide vane of water-jet assembly, which needs to be optimized.

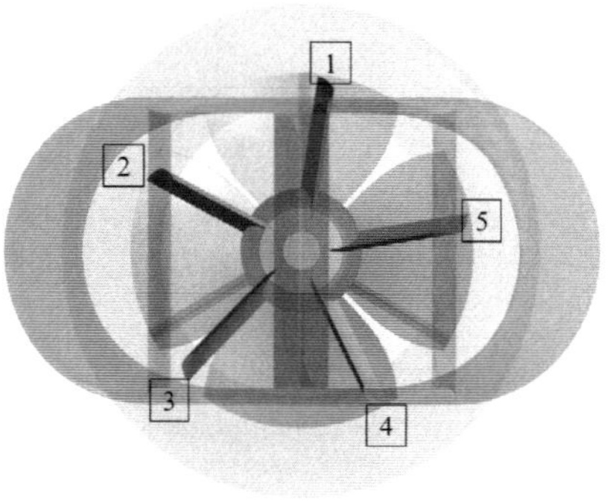

Figure 9. Guide vane position

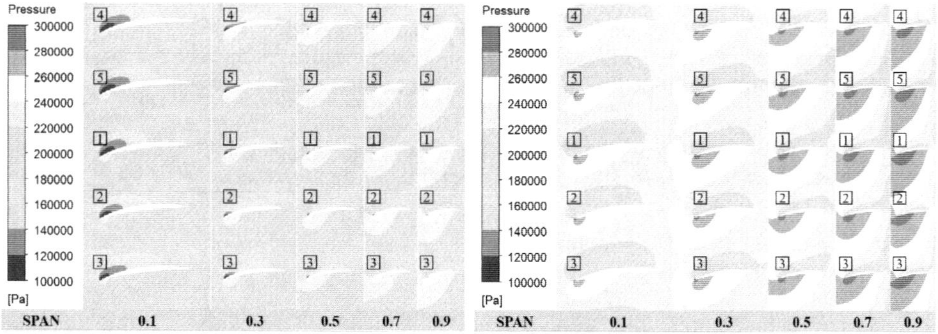

Figure 10. Pressure distribution in guide vane

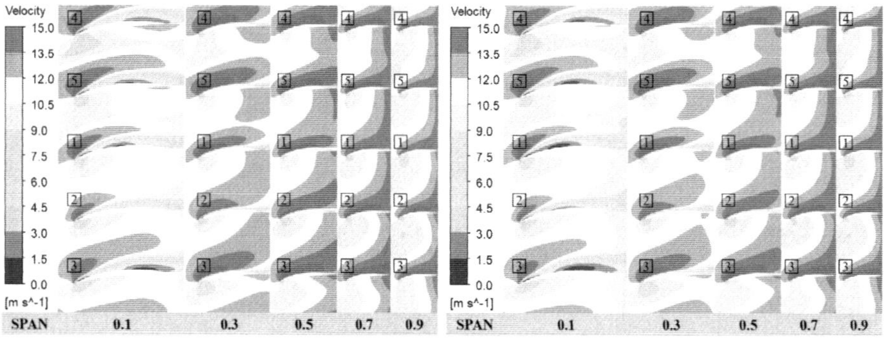

Figure 11. Velocity distribution in guide vane

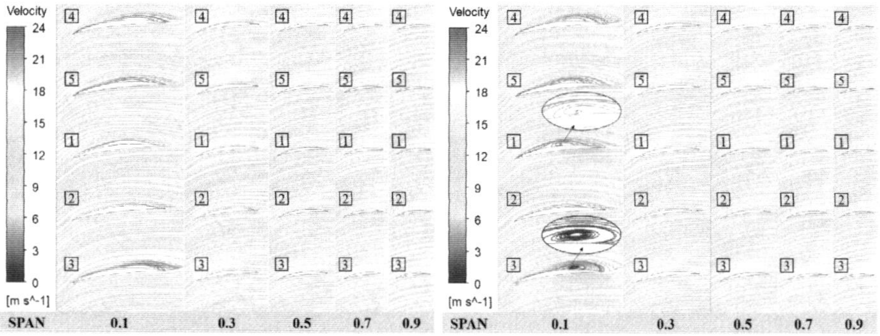

Figure 12. Streamline distribution in guide vane

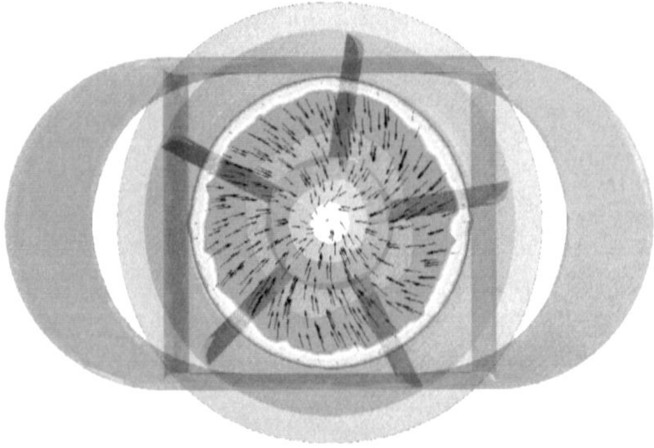

Figure 13. Circumferential vector distribution at guide vane outlet

4. CONCLUSIONS

Through the numerical simulation of direct navigation and reverse navigation of water-jet assembly, the following findings are found:

(1) During reserve state, the middle rudder plate and rudder shell form a closed cavity. The pressure rise is caused by the blockage of the flow channel. The pressure rise area extends from the closed cavity to the impeller, and the jet at the nozzle is sprayed reversely from the gap to realize the reverse sailing of the ship. However, the reverse jet is different from the original design intention.

(2) During reverse state, the overall pressure in the guide vane increases, but the velocity distribution has little difference. Due to the increase of outlet pressure, the reverse pressure gradient near the suction surface of the guide vane increases and the velocity decreases in the area close to the hub, and finally the reflux vortex is generated at the turn of the arc surface of the suction surface.

(3) Due to the different inlet angles of fluid at different span of guide vane, but the angles on different spans of the blade are consistent, the fluid near shroud still has a certain circumferential velocity after flowing through the guide vane.

Only the steady numerical calculation of water-jet assembly is carried out this time. The influence of impeller rotation position on the flow characteristics of water-jet assembly is not known. Unsteady calculation will be carried out for more in-depth research in the future.

5. ACKNOWLEDGEMENTS

This research is funded by National Youth Natural Science Foundation of China (51906085). China Postdoctoral Science Foundation Funded Project (2019M651734). Jiangsu Province Innovation and Entrepreneurship Doctor Project (2019). Zhejiang Postdoctoral Project (2019). The NSFC (Natural Science Foundation of China) (U20A20292). Project of State Key Laboratory of mechanical system and vibration (MSV202203).

6. REFERENCE

1. WANG L., 'Water-jet propulsion and water-jet propulsion pump', *General machinery*, 2007.

2. TIAN N, CHEN J, WANG L., 'Development and application perspective of an 1000kW Waterjet-Assembly'. *Ship and Boat*. 2007.

3. CHANG S, LI K., 'Effects of Leading Edge and Trailing Edge Design on Performances of Waterjet Pump'. *Ship Engineering*, 2019.

4. ZHANG Y, WANG L, ZHONG J, et al., 'Influence of Multi-parameter Optimization Design on Hydraulic Performance of Water-jet Propulsion Assembly'. *Journal of drainage and irrigation machinery engineering (JDIME)*, 2021.

5. HAN W, XU D, GUO W, et al., 'Effects of Axial Clearance on Hydraulic Characteristics of Counter-Rotating Water-Jet Pump', *Journal of Propulsion Technology*, 2019.

6. GONG J, GUO C, WU T, et al., 'Numerical Study on the Unsteady Hydrodynamic Performance of a Waterjet Impeller'. *Journal of Coastal Research*, 2018.

7. LU Y, WANG X, LIU H, et al., 'Analysis of the impeller's tip clearance size on the operating characteristics of the underwater vehicle with a new high-speed water jet propulsion pump'. *Ocean Engineering*, 2021.

8. WU H, MIORINI R L, KATZ J., 'Measurements of the tip leakage vortex structures and turbulence in the meridional plane of an axial water-jet pump'. *Experiments in Fluids*, 2010.

9. XU S, LONG X-P, JI B, et al., 'Vortex dynamic characteristics of unsteady tip clearance cavitation in a waterjet pump determined with different vortex identification methods'. *Journal of Mechanical Science and Technology*, 2019.

10. MIORINI R L, WU H, KATZ J. 'The Internal Structure of the Tip Leakage Vortex Within the Rotor of an Axial Waterjet Pump'. *Journal of Turbomachinery*, 2012.

11. ZHANG W, SU S, Feng Xuedong, et al., 'Effect of the water inlet area on characteristics of waterjet duct'. *Journal of Jiangsu University of Science and Technology (Natural Science Edition)*, 2016.

12. SHI J, FENG X, LI G, et al., 'The Calculation Method of Shafting Torsional Vibration with Ice Impact'. *Ship and Ocean Engineering*, 2016.

13. JI G, CAI Y, LI N, et al., 'Influence of Lip Parameters on Non-Uniformity and Stagnation Point at Inlet Duct of Waterjet Propulsion', *Shipbuilding of China*, 2016.

14. WANG J, LI G, ZHANG Y, et al., 'Research of forces on rudder of integrated waterjet steering and reversing gear based on CFD technology'. *Chinese Journal of Hydrodynamics*, 2022.

15. LI G, WANG J, WANG L Wang., 'Research of forces on the steering and reversing gear of waterjet'. *Chinese Journal of Hydrodynamics*, 2020.

16. LIU X. 'Research on Structure Strength and Stabilization of the Built-up Rudder of Waterjet Assembly'. *Harbin Engineering University*, 2012.

17. YU H, DUAN N, HUA H, et al., 'Propulsion performance and unsteady forces of a pump-jet propulsor with different pre-swirl stator parameters'. *Applied Ocean Research*, 2020.

18. CEN C., 'Study on Flow Characteristics and Optimization Design of Water-jet Propulsion Reversing Bucket'. *Jiangsu University*, 2019.

AUTHORS BIOGRAPHY

Yun Long holds the current position of associate researcher at National Research Center of Pumps of Jiangsu University. He is responsible for the study of the internal flow of fluid machines. His previous experience includes the research on hydrodynamics of high velocity regions in a water-jet pump based on experimental and numerical calculations at different cavitation conditions, etc.

ROUTE OPTIMISATION FOR ADVANCED WATERJET SELECTION

A Tomkins and **D Muller,** HamiltonJet, New Zealand

SUMMARY

Total vessel efficiency is about much more than the top speed performance. Whilst top speed performance plays a part in the overall efficiency of the vessel, taking a more holistic approach to the operation cycle of vessels can yield significant gains in efficiency. Designing waterjets that can offer performance advantages across a vessels complete speed range can be a challenge as it often requires finding a balance between competing demands.

However, by undertaking detailed route analysis of our customers operations, HamiltonJet are able to assess each phase of a vessel's operation and tailor equipment selection and waterjet design to deliver optimal all-round performance.

This paper will cover the methodology used to undertake these studies and share some of the interesting insights and developments in product and waterjet selection methodology that have been discovered through this work.

NOMENCLATURE

A nomenclature is required for papers/technical notes using a large number of symbols, abbreviations and acronyms. SI units should be used wherever possible and ordered alphabetically.

ESS	Energy Storage System
ICE	Internal Combustion Engine
PTI	Power Take In
PTO	Power Take Out
EM	Electric Machine
[Symbol]	[Definition] [(Unit)]
u	Kinematic viscosity (N s m^{-2})
r	Density of water (kg m^{-3})
P	Pressure (N m^{-2})

1. INTRODUCTION

Total vessel efficiency is made up of much more than a value at a single point of sailing and, whilst it may be easy for naval architects and industry suppliers to look at one key metric such as ultimate vessel speed, this can often provide a misleading picture as to the vessels true all-round efficiency.

In today's world of deep focus on energy efficiency important gains can be made by taking a more holistic look at a vessels operation cycle by taking into account all aspects of the vessel's operation from docking through to high-speed transits.

Finding the optimal balance of performance and efficiency is often a balance of compromises between outright speed and lower speed thrust and manoeuvrability.

By considering these aspects in detail, HamiltonJet has been able to optimise vessel jet selection, and new jet designs to give operators significant gains in vessel efficiency, and in turn open up more applications to waterjet propulsion, particularly those applications which may not have historically been thought to be suitable for the system.

This approach is directly impacting how we now select water jets and develop our new products, furthermore, as the industry moves towards low and zero emissions vessels this approach is critical for ensuring that propulsion solutions are well optimised and fit for purpose. Failing to consider the complete operation cycle in these low and zero emissions applications can lead to applications which considerably miss their desired performance goals.

This paper and presentation will outline the methods, results and conclusions gained from undertaking this work based around two main case studies.

2. CASE STUDY 1 – 40M HIGH SPEED FERRY

2.1 OVERVIEW

The first study we will look into concerns a high-speed ferry service running in Europe. The operators run a 30-minute-high speed commuter service between the mainland and an island with services running throughout the day. The operator was looking to procure a new vessel to increase the efficiency and reliability of the daily service and invited HamiltonJet to explore potential efficiency savings that could be achieved by considering an alternative propulsion system to the existing fleet.

The traditional approach to "increasing" efficiency for such an operation type is to seek a higher vessel speed for the same installed power. However, whilst this is a measure of the ultimate propulsive efficiency of the propulsion system this doesn't accurately reflect true total efficiency of the vessel over the whole operation profile as it is only concentrating on one, potentially small, part of the operation profile.

To gain a greater understanding of the total efficiency of the vessels propulsion line it is important to look at each stage of the vessel's operation profile. For each vessel operation type the operation profile will be different but for a ferry service the operation can be split into three main modes:

- Transit phase
- Acceleration phase
- Docking phase

If we take each phase in isolation and study how differences in propulsion systems can impact the outcome, we can then build a detailed picture of the total efficiency of the vessel, and, more importantly, understand how alternative waterjet selections can alter the total efficiency of the vessel.

If we look at each of the phases in isolation, we can see how they have differing demands from the propulsion system.

Transit phase. This is the high-speed section of the operation profile, and the phase which requires top speed of the vessel to be taken into account. For waterjet selection high speed efficiency is generally achieved with a smaller relative diameter in relation to the input power as the smaller jet creates more favourable inlet velocity ratio (IVR) conditions at higher vessel speeds.

Acceleration phase. This phase is essentially the transition mode of the vessels operation cycle as it moves from the docking phase into the transit phase. Whilst often overlooked, if this phase is managed to be shortened due to high accelerating propulsors, good efficiency gains can be made by allowing to the vessel to run its optimal transit state for longer, or by allowing the transit speed to be reduced and therefore saving energy. Increases in acceleration for waterjet vessels are general achieved by increase the thrust margins delivered by the jet which can be achieved by either increasing the cavitation resistance of the jet, and thus allowing higher power to applied sooner, or by increasing the diameter of the waterjet and increasing the net thrust available

Docking phase. This area can be critical to the overall efficiency of the vessel. Even for high-speed vessels on reasonably long runs the docking phases of the operational cycle can account for a surprisingly high percentage of the total operation. In our particular case study this phase accounted for 25% of the total sailing time. It is clear to see from this that any gains here which allow the vessel to dock faster with lower power can make a significant impact on the vessel's efficiency. This particular phase is important when considering waterjets versus other propulsions systems, even for slower speed vessels, as the waterjets superior manoeuvrability can dramatically swing the total efficiency of the vessel in its favour of waterjet-based systems.

2.2 MODELLING

The first step in the process is to gain a detailed understanding of the operation profile of the vessel. For existing operations this can be relatively easily achieved by data logging existing vessels running the same routes. This can be done either by dedicated onboard data loggers or by accessing high resolution AIS data.

For our case study our start point was by using high resolution AIS data points. By accessing 3 months of data records for existing vessels running the same route during the vessels' high season we were able to analysis the routes and build a model of an average sailing profile in both outbound and return directions. It is important to take a large sample size of the AIS data to minimise any variability caused by weather conditions, crew variances and unexpected vessel movements due to traffic etc.

This route information gives us an accurate picture of vessel speeds, course and heading at regular intervals on the vessels journey and gives us a detailed operation profile of the vessel.

The next step is to model the behaviour of the vessel itself. For this we undertake a performance estimate for the vessel. This is a common step in the jet selection process being the basis of any waterjet selection. This is done by modelling the vessels resistance, either by in house calculations or by having these supplied from CFD or tank test data. Jet thrust data for the given installed power is then overlaid.

Unlike a regular performance estimate where only a few data points are analysed, for route analysis performance estimates we analyse many more speed points – the higher the number the greater the fidelity of the final analysis. For this study we took data points for each knot of vessel speed which told us the vessel resistance, thrust required to maintain the speed, and the power needed to generate that thrust from the waterjet. This gives us a detailed powering requirement for each speed point of the vessel's operation. From this and using the engine manufacturers specific fuel consumption values for the given engine power and rpm we derive the fuel burn at each of these speed points and start to consider the optimisation possibilities.

2.3 OPTIMISATION

The first stage in the optimisation process is to look at the total route profile and assess whether there any potential propulsive efficiency (PE) gains to be made by altering the profile of the route.

In this case study the percentage of time that the vessel spends at top speed (59%) initially leads thinking towards focusing on the high-speed efficiency of the vessel. However, at these speeds' efficiency differences between different propulsion models and designs can be relatively small at 1-3%. Over a relatively short sailing profile such as this one, the potential PE gains of the transit phase are marginal.

The remaining 41% of the operation profile is made up of the docking phases and acceleration phases so these zones of operation must also be taken into account if we are going to consider the total operational efficiency of the vessel. Equally these are also the phases in which passenger vessel operators experience most delays.

Medium and low speed thrust can be significantly impacted by jet design and model selection so for these phases its critical to assess the impact of an alternative options.

For our study we began by considering a waterjet selection which focused less on the ultimate top speed of the vessel (although this is still important) and more on an all-round balance of performance.

Putting this jet model into our route model delivered some interesting results. Top speed was reduced about around 1.5 kts, however, thrust available during the docking phases was significantly higher – in the region of 50% at the lowest speed points. This gain means that it is possible for the vessel to either dock faster (using the same power level) or maintain the same profile using lower power inputs and lower corresponding fuel burn.

For the acceleration phase, the higher thrust margins afforded by the alternative jet selection enables greater thrust for the same power input and thus helps to increase acceleration and reduced the time spent in the acceleration phase, however, this effect does lessen as the vessel approaches higher speeds.

Applying this jet selection to the modelled route profile delivers some interesting results and offers some alternative operation profiles for the operator.

The biggest gains are apparent in the docking phases so this is where most of the optimisation can be made. This gives us two new routing options.

Route A. This profile has the same top speed during the transit phase, and similar docking speeds but utilising less power at during this phase.

Route B. This profile exploits more of the gains in the docking phases by having greater propulsive force and therefore reducing the time taken to dock. This allows the same total sailing time, but a lower top speed is then required in the transit phase.

Route profile A provides some improvement in overall efficiency as it enables some fuel savings in the docking phases, however, as the transit speeds are not altered the over increase in efficiency is smaller than other optimisations.

Route B presents a much greater opportunity for improving efficiency. By modelling the docking phase we can calculate that the time to dock the vessel can be reduced by up to 50% through increased thrust margins giving greater propulsive force for the same power, and by highly efficient control surfaces delivering near lossless vectoring of the jet thrust.

As the docking phase on this particular sailing profile represents around 25% of the total sailing time improvements of this magnitude in this phase can have a significant impact on the overall efficiency of the vessel. Greater manoeuvrability allows the vessel the to dock faster with lower power inputs which in itself delivers some fuel saving, however, the greatest gains come when we then consider how this affects the rest of the operation profile.

By reducing the docking time, the vessel can now travel at a lower transit speed whilst still maintaining the same total sailing time, this allows the main engines to run at lower rpm and further fuel savings can be achieved.

For this study the difference between the standard jet selection and the optimised one is significant given the relatively short transit time.

	Standard route	Optimised route (B)
Vessel	40m catamaran	40m catamaran
Waterjet	Jet for top speed	Jet for optimised profile
Driveline	4 x 900kW @ 2,100rpm	4 x 900kW @ 2,100rpm
Max vessel speed (kts)	41	40
Required transit speed (kts)	38	31
Sailing time (minutes)	27	27
Fuel consumed (l)	212	188

By optimising for the total sailing profile rather than purely for top speed the operator can gain a fuel saving of around 11% with the only sacrifice being around 1kts in vessel speed. Additional gains come from the fact that the operator still has the option to run at the higher transit speeds if required which will allow for further flexibility in the service and enable the operator to recover any lost time to delays.

The majority of the gains seen here are delivered by selecting a waterjet that is larger in diameter for the relative power, but a balance needs to be struck between sizing for the low-speed operation and high-speed operation. Focusing solely on the low-speed operation by "oversizing" the water jet will have too great a negative an impact on the top speed, however, by undertaking this route analysis we can get new insights into how to find the optimal selection for the complete operation cycle, and in turn by challenging what is considered as "efficiency" this opens up more applications to waterjets, which perhaps traditionally would have been considered not suitable for waterjets.

3. CASE STUDY 2

We have seen how the route analysis tools can help optimise conventional powered vessels to increase the overall efficiency, however, these techniques take on even greater significance when low and zero emission vessels. For these complex applications developing a deep understanding of the power and energy demands of the vessel can make the difference between vessels that perform well and deliver true low emission advantages and those that fail to meet their goals.

The case is for a 32m passenger catamaran. The main goal of this case is to reduce the use of fossil fuels as much as possible while maintaining the ability to cater for several different routes with speeds up to 26 knots, carrying 300 passengers, with a limit of 3MW of charging capacity at one terminal.

The best way to achieve the goal would be to use a full-electric solution; however, at the time of the design, the energy density of marine batteries imposed a limitation for a full-electric vessel as their weight and charging times would not allow operation on the longest routes. For that reason, the chosen solution was to design a hybrid-fuel propulsion that used as much electric energy stored in batteries as possible, complemented by Diesel engine power. It is understood that in the future, when energy storage systems (ESS) become more energy-dense and able to charge faster, that a full electric conversion may be considered. The process of designing the right system involved studying and optimising several configurations of waterjets, motors, engines, battery pack, series and parallel hybrid options and evaluating their performance over a range of routes and several different displacements.

3.1 EHX (electro-hybrid system)

EHX (electro-hybrid system) is a HamiltonJet proprietary propulsive power management system built upon AVX (advanced waterjet control system) specifically designed to control the propulsive energy flow of marine vessels. It manages the flow of energy into and out of the ESS into and out of the electric machines (EM), out of the internal combustion engines (ICE) and into the waterjets. EHX also controls, gearbox clutch engagement and engine start/stop. It can be used in parallel hybrid, series hybrid and full electric vessels to operate in the several different modes of operation that these systems allow such as electric only, ICE only, ICE generation, hybrid, or boost. The transition between modes can be done automatically as a function of throttle or rpm input or manually via buttons. Multiple parameters can be configured in EHX such as thresholds for mode change in automatic mode (e.g. variables like battery state of charge (SOC), rpm and time), sequence of modes and pre-sets that hold different sets of variables. Since EHX controls the entire propulsion system and allows flexibility in programming modes, it enables the fine tuning of the operation of the vessel, and optimized sizing of propulsive equipment.

3.2 SERIES VS PARALLEL HYBRID

Two main alternatives were considered for hybrid propulsion of this passenger ferry. One, in parallel, where power is delivered mechanically to the waterjet by both an ICE and an electric motor, and another, in series, where the waterjet receives mechanical power only from an electric motor. In both cases, batteries are used as a propulsive ESS.

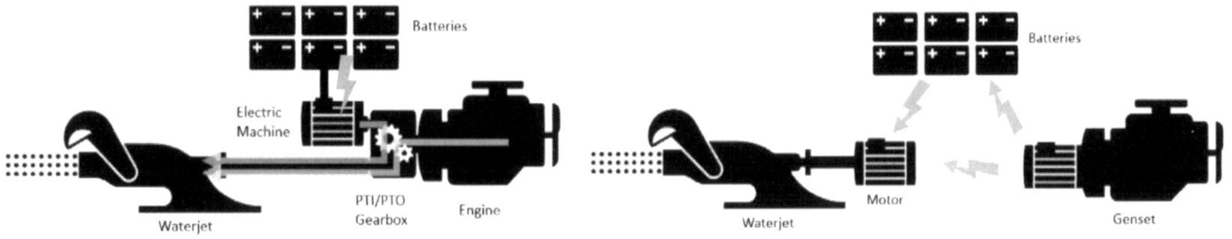

Figure 1: Schematics. Left: Parallel Hybrid System with ESS. Right: Series Hybrid System with ESS.

Although generation mode (charging the ESS using ICE) is possible in both parallel and series hybrids, this mode is available only for emergencies in this vessel, since the main objective is to use as much electricity from the grid as possible to charge the batteries rather than using the diesel engines or gensets.

3.2.1 Parallel Hybrid

In the case of parallel hybrid, a PTI/PTO gearbox was selected as the means to allow power from the ICE and/or the motor to reach the waterjet (PTI). This gearbox can also be used to divert power from the ICE to charge the batteries using the motor as a generator. Since the motor can be used both as motor and generator, it is often referred to as an electric machine (EM). There are five possible modes of operation in parallel hybrid: diesel only, electric only, generation, hybrid, and boost. Figure 2 illustrates these five modes in as schematics showing the energy flow as arrows. Note that hybrid and boost modes work in a similar manner. Boost mode generally refers the added electric power after the ICE has reached its full potential. Hybrid mode related to any ratio of electric/ICE power that can be obtained at medium rpms, without either the EM or the engine reaching full power.

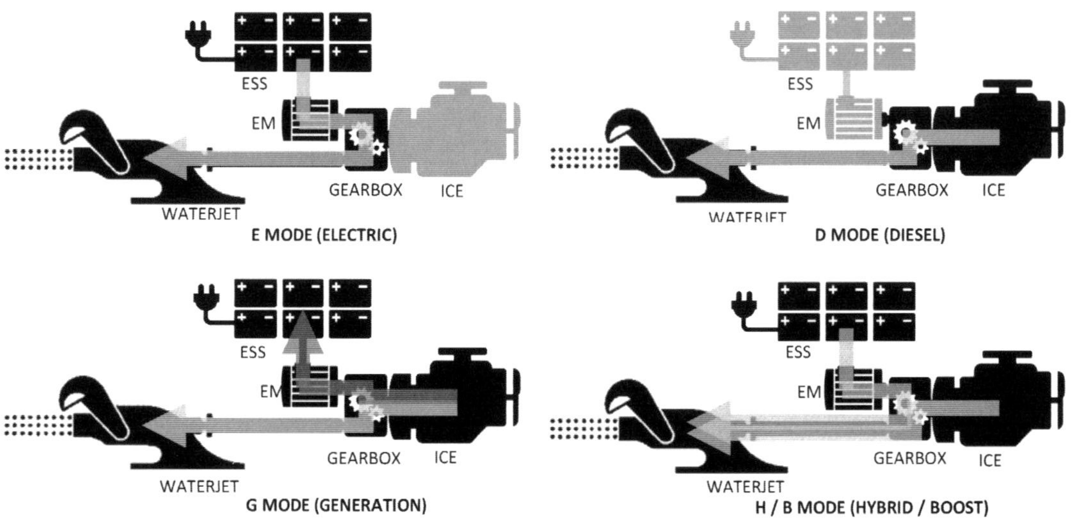

Figure 2: Different modes of operation in parallel hybrid.

The EM in series hybrid systems is normally much smaller than the ICE in terms of power output and is used in to add the ability to operate in electric mode for short periods and at low speeds, with a limited reduction in the use of diesel. However, in this vessel, due to the goal of reducing fossil fuel consumption, the EM and ICE were sized to deliver similar power, with the EM sized at 85% of the power of the ICE. To achieve the maximum speed, boost mode is required by adding up the full power from the EM and ICE. The maximum speed achievable in electric only mode is of the order of 14 knots.

For operation in automatic mode, EHX is programmed to transition from electric directly into hybrid mode as a certain rpm threshold is reached. Figure 2 shows the power curves of the ICE, EM and waterjet as function of rpm, and the power and how the power is drawn from the EM and ICE (also as a function of rpm).

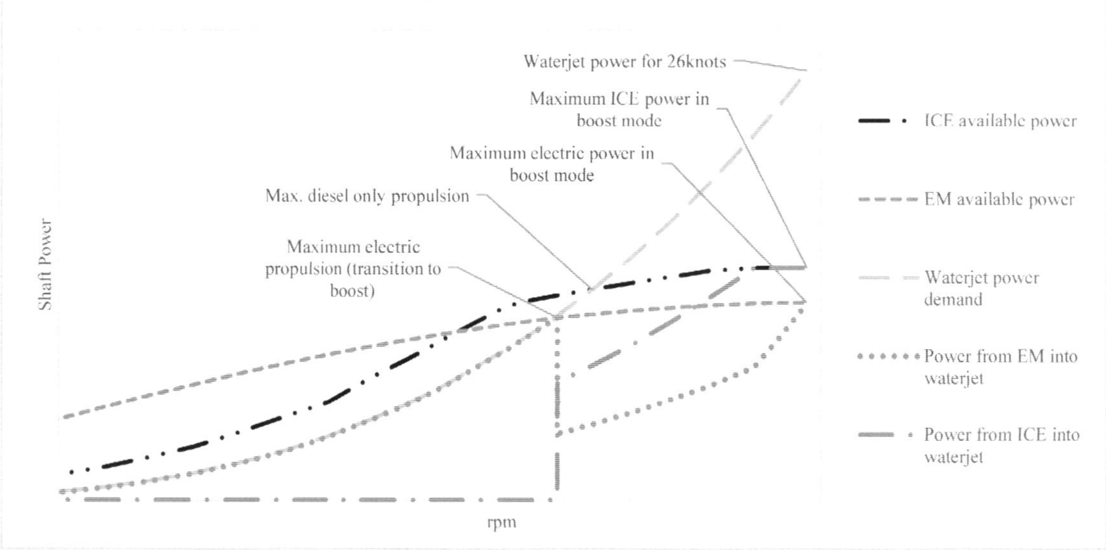

Figure 3: Power map and automatic mode transition for parallel hybrid option.

3.2.2 Series Hybrid

In a series hybrid vessel, the electric motor delivers all the power required by the waterjet to reach maximum speed. In this vessel, the electric power is provided by the ESS and two diesel gensets, allowing five main modes of operation: full electric (ESS only), generation (batteries charged by the gensets), hybrid (waterjets powered by both gensets and ESS), gensets only and charging only. Intermediate modes of operation can be added, such as single genset and dual genset operation. Figure 3 shows all these modes in schematics. The batteries were sized so that the ESS alone can supply al the power to the waterjets to reach top speed, and the diesel gensets are used as range extenders to allow sailing to the more distant routes. The gensets on their own can only achieve approximately 12 knots.

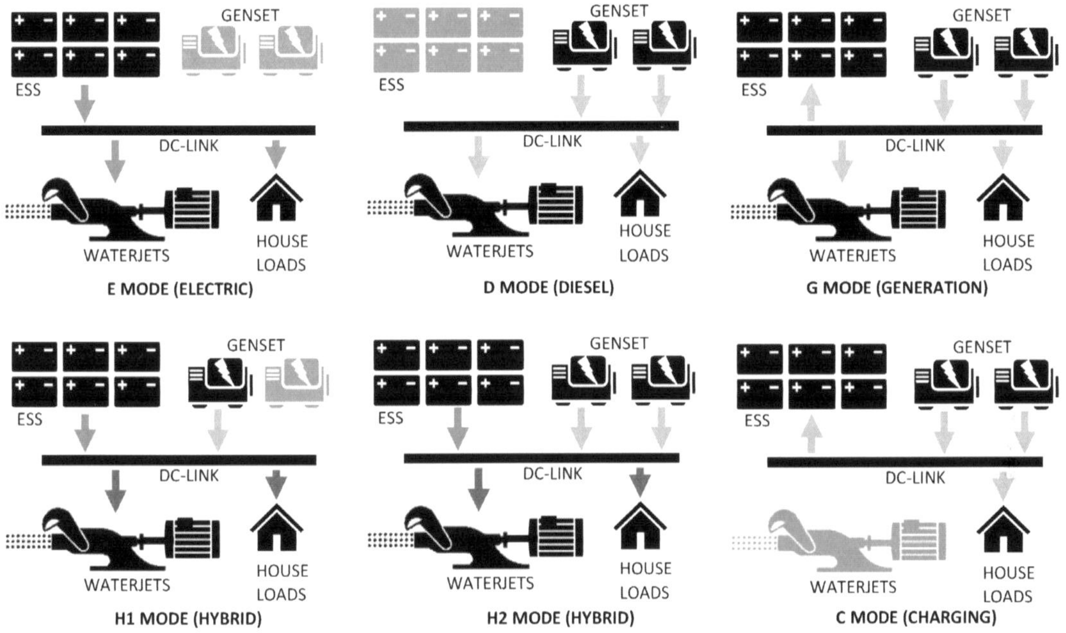

Figure 4: Six possible modes of operation in the Series Hybrid configuration.

Although the vessel can reach full speed on ESS alone and operate as a full electric vessel for short distances, in longer routes it can only travel for a limited distance on electric mode before the gensets need to be used, which would force a reduction in speed. If full speed is to be maintained for longer distances, the vessel needs to operate in hybrid mode from the start of the journey. Therefore, the skipper needs to activate the appropriate mode at or shortly after departure. Certain routes may require the gensets to operate at partial load, only for a certain period or at certain speeds. This means that different routes require different pre-sets which can be programmed into EHX. Figure 3 shows four possible configurations (pre-sets) for operation on automatic mode that can be selected for different routes depending on distance. The mode changes according to rpm. All configurations start in electric mode for manoeuvring and low-speed operation, adding diesel genset power at higher rpms in higher quantities as needed to cover longer distances. In these examples, the gensets are either at 100% capacity or off completely. However, any intermediate load percentage can be programmed and stored as a pre-set, allowing for precise adjustments so that the minimum amount of fuel is consumed in any specific route. The fourth pre-set in Figure 3 includes power generation (G mode), meaning that the gensets provide energy for propulsion, house loads and battery charging at medium rpms.

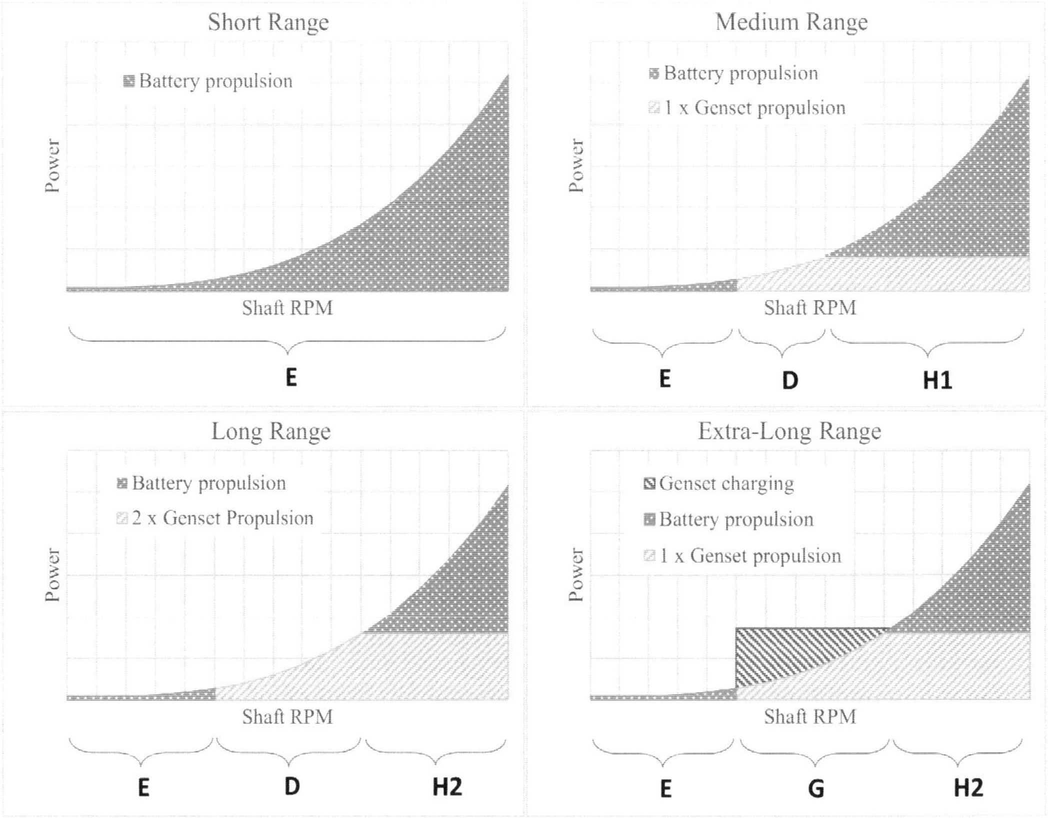

Figure 5: Possible configurations of automatic modes depending for various rout distances as function of rpm.

3.2.3 System selection

When gensets are used to propel the vessel, there are significant losses associated with the conversion of mechanical power at the ICE into electricity at the generator and back into mechanical power at the motor to power the waterjet. These losses which are of 15 to 20% are not present in a parallel hybrid system. Moreover, in a parallel hybrid system the same EM is used as generator and motor, whereas in the series hybrid these are separate, and the motor is larger, resulting in additional weight. Despite the heavier installation, series hybrid can operate at full speed as a full electric vessel for short routes, whereas the parallel hybrid always required ICE power for high speeds. There are other advantages and disadvantages of each system regarding the operation of this ferry which are listed below. However, one crucial aspect of the series hybrid system led to its selection for implementation: the ability to convert it to a full electric vessel in the future when better ESS technology becomes available and the gensets can be eliminated, which may even coincide with the end of life of the batteries installed.

	Advantages	Disadvantages
Parallel hybrid	• Single EM as motor and generator • Lighter installation • Lower capital investment • Power directly delivered mechanically from both ICE and EM to the waterjets	• Requires PTI/PTO gearbox • Generation mode only available over a small range or rpms • Limited speed in electric only mode • More difficult to convert to full electric in the future
Series hybrid	• Full electric mode available at all speeds • No PTI/PTO gearboxes required • Generation available at any rpm • Easier to convert to full electric in the future	• Separate motors and generators • Conversion from mechanical power at the ICE to electric power at the motors incurs inefficiency • Heavier installation • Higher cost

3.3 WATERJET SELECTION

For the parallel hybrid design, a twin waterjet configuration was selected. The main reason to have a single propulsion line in each demi-hull is the size of the ICE, EM and PTI/PTO gearbox package in the engine room, which would make any multiple-waterjet configuration difficult to accommodate. Figure 3 illustrates schematically how the propulsion lines are accommodated in the hull.

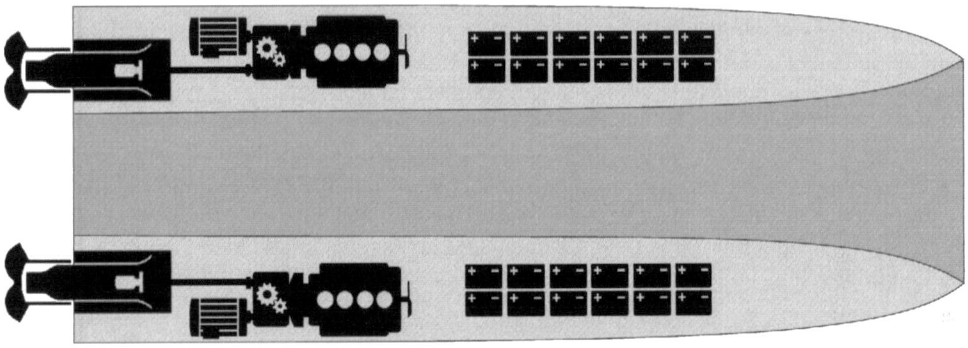

Figure 6: Parallel hybrid propulsion schematic

The waterjet model selection in this case depends on several factors: its size, weight, power capacity, cavitation limits, efficiency, and rpm range. Reducing weight is an important consideration in the design as it has a direct impact in the hull resistance and, therefore in the sizing of the ESS. The efficiency of the waterjet also has an impact in the size and weight of the ESS. However, waterjets of similar hydrodynamic characteristics, but different sizes, have better efficiency the larger they are, and the additional weight of a more efficient waterjet has a detrimental effect in the hull resistance. In order to determine the ideal configuration, four potential waterjet models were considered based on power capacity and size: HTX52, HM721, HM811, and HT810. Figure 2 shows the comparison between different waterjets in weight, and the impact of the efficiency in the weight of the ESS.

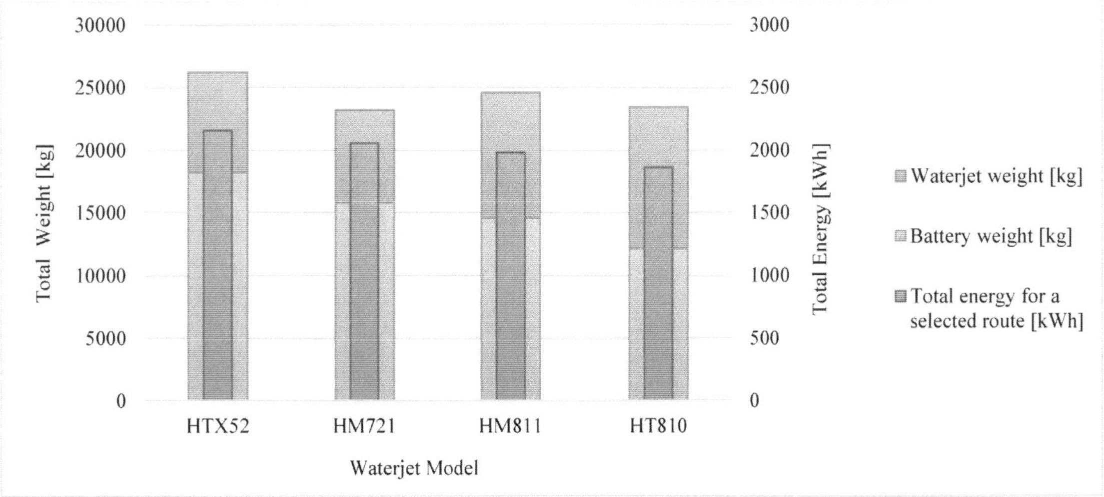

Figure 7: Weight comparison for different waterjet models from the smallest on the left to largest in the right.

Although the HM721 waterjet represented the lowest-weight case, the HT810 waterjet as a close second requires less energy to cover the same distance for a specific route, therefore consuming less fuel and offering a faster charging time. However, during the subsequent vessel detail design, the HT810 waterjet no longer could be fit in the space available in the hull and the HM721 model was selected as the best option for a parallel hybrid solution.

For the series hybrid option, two configurations were considered to select a waterjet model: a twin configuration and a quad configuration. The twin configuration required large electric motors with low rpms. These are only available as bespoke builds, are costly and have long lead-times. Alternatively, a smaller but higher revving motor could be used if a gearbox were employed; however, there are mechanical inefficiencies and extra weight associated with that. A quad-waterjet configuration utilising the largest waterjet that could be fitted in the demi-hulls can be powered by smaller motors that operate at more favourable rpms without requiring gearboxes. Also, the simplicity and compactness of electric motors directly driving waterjets enables the use of two propulsion lines per demi-hull, unlike the parallel hybrid option. Figure 4 shows a schematic representation of this propulsion arrangement. These factors led to the selection of the HTX42 waterjet model.

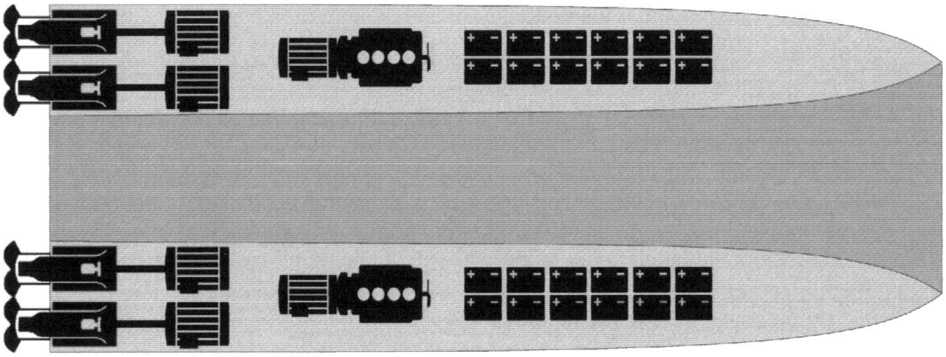

Figure 8: Series hybrid propulsion schematic

4. CONCLUSIONS

Optimising waterjet selection to achieve high all-round efficiency for the vessel requires a far more considered approach to the application engineering process than traditional propulsion selection processes. As we have seen adopting this more holistic approach to the vessel's operation profile can help to deliver vessels that are more efficient, lower emission and more successful in meeting customer expectations.

These techniques can be applied with good success in the application of next generation low and zero emission vessels but also for vessel applications that may not traditionally be seen as waterjet applications opening up the benefits of waterjet propulsion to a wider user group.

A METHOD TO EVALUATE THE PERFORMANCE OF WATERJET PROPULSORS USING A MODEL SCALE SETUP

KGDL Dias and **SASBS Suraweera**, Walkers Shipyards, Sri Lanka
TS Ruwandi Ocean University, Sri Lanka

SUMMARY

Waterjet propulsor is a key component in the Small High-Speed Craft (SHSC) industry. Although the SHSC industry is well established in Sri Lanka, no waterjets are designed or manufactured due to the absence of prototype testing facilities in the country. The availability of such facilities provides the opportunity not only for the waterjet designers and manufacturers but also for the shipyards and design offices to evaluate the performance of waterjets when selecting propulsion systems at the preliminary design phase of their craft.

Therefore, a method is proposed to test and evaluate the performance of waterjets using a model scale test setup in the early design stage for a Small High-Speed Craft. In this setup, a waterjet mounted on an SHSC- hull is fully submerged and secured in a water tank with a circulation pump loop setup. The waterjet is driven by an electric motor under varying flow conditions and the performance is evaluated using different measures, such as the head, power, torque and efficiency as recommended in the existing methods.

This paper presents a method to appropriately scale the entire test setup, select a suitable hull-jet configuration and proposes an approach to evaluate waterjet performance with the aid of existing measures and methods using the proposed model scale test setup.

NOMENCLATURE

ρ	- Density of water (kg/m^3)
P	- Pressure (N/m^2)
W	- Displacement Weight
τ	- Trim angle of planning area (deg)
C_{F0}	- Friction Coefficient (ITTC 1957) or ATTC (Function of $Rn_b = \frac{V_1 \lambda b}{\nu}$)
β	- Dead-rise angle at mid-chine position (deg)
λ_m	- Mean wetted length beam (m)
A_n	- Nozzle area (m^2)
V_o	- Outlet velocity (m/s)
V_i	- Inlet velocity (m/s)
V_s	- Design speed (m/s)
λ	- Scale ratio
V_m	- Velocity at model scale (m/s)
Q / Q_J	- Flowrate (m^3/s)
τ	- Torque (Nm)
ω	- Angular Velocity (rad/s)
n	- No. of Revolutions per second
Q_{shaft}	- Torque of the shaft (Nm)
IVR	- Inlet Velocity Ratio

1. INTRODUCTION

Waterjet propulsor is a key component in the Small High-Speed Craft (SHSC) industry [1-3]. This propulsor offers many advantages over other types of propulsors. It is inherently more efficient than conventional propellers at speeds of over 25 knots. In addition, excellent manoeuvrability, minimal noise and vibration, operation at limited drafts, increased safety, lower drag are all compelling features which promulgates the usage of waterjets in the SHSC industry.

This industry is well established in Sri Lanka with many manufacturers focusing on both the local and international markets catering to navies, coast guards, rescue agencies including the recreational market. However, no waterjets are designed or manufactured due to the absence of prototype testing facilities in the country. The availability of such facilities provides the opportunity not only for the waterjet designers and manufacturers but also for the shipyards and design offices to evaluate the performance of waterjets when selecting propulsion systems at the preliminary design phase of their craft [4-9].

Therefore, a method is proposed to test and evaluate the performance of waterjets using a model scale test setup in the early design stage. In this setup, a waterjet mounted on an SHSC- hull is fully submerged and secured in a water tank with a circulation pump loop setup. The waterjet is driven by an electric motor under varying flow conditions and the performance is evaluated using different measures, such as the head, power, torque and efficiency as recommended in the existing methods [10].

Several methods have been recommended in the previous studies to evaluate the performance of waterjets. The International Towing Tank Conference (ITTC) recommended procedure is to evaluate the propulsive performance of a waterjet driven high speed craft using the resistance and self-propulsion test as described in [11] and evaluate the propulsor performance separately of the waterjet system using a cavitation tunnel or pump loop setup as described in [10]. Although this method based on the momentum flux approach is recommended, ITTC acknowledges that further research pertaining to the direct thrust method as described in [12, 13] is deemed necessary. In addition to experimental methods, numerical methods based on CFD are already in practice to evaluate waterjet performance as described in [14, 15]. However, the proposed method combines both the experimental and numerical methods to evaluate waterjet performance in a model scale setup.

This paper presents a method to appropriately scale the entire test setup and to select a suitable hull-jet configuration. In addition, a method for evaluating waterjet performance is described with the aid of existing measures and methods using the proposed model scale test setup.

2. METHODS

A method is developed to test and evaluate the performance of waterjets using a model scale test setup consisting of three major parts viz. hull, waterjet and testing tank The method is developed in four consecutive steps a) hull b) waterjet c) scaling and d) testing tank and detailed under the following sub-sections.

2.1 HULL

In this study, SHSC hulls are of main interest. They are less than 24m in length with a Froude number greater than 1.1 as defined in [16, 17].

First, the resistance curve is computed for the design-speed curve of the selected SHSC using Savitsky method as described in [18, 19] using

$$R_{Ts} = \mathrm{W}\tan\tau + \frac{1}{2}\rho V^2\lambda_m b^2 C_{F0}/(\cos\tau\cos\beta) \tag{1}$$

where,

R_{Ts} - Resistance of the vessel [N]
W - Displacement weight [kg]
τ - Trim angle in [degrees]
ρ - Density of water [kg/m^3]
V - Design speed of the craft [m/s]
λ_m - Mean wetted length-beam [m]
b - Beam of the planing area [m]
C_{F0} - Friction coefficient according to the ITTC 1957 or the ATTC friction lines as a function of $Rn_b = \frac{V_1\lambda b}{\nu}$,
β - Dead-rise angle at the mid-chine position [degrees].

The effective power (P_E) of the craft is then calculated by

$$P_E = R_{Ts} x\, V_S \tag{2}$$

to obtain the corresponding power curve, where V_S is the design speed.

2.2 WATERJET

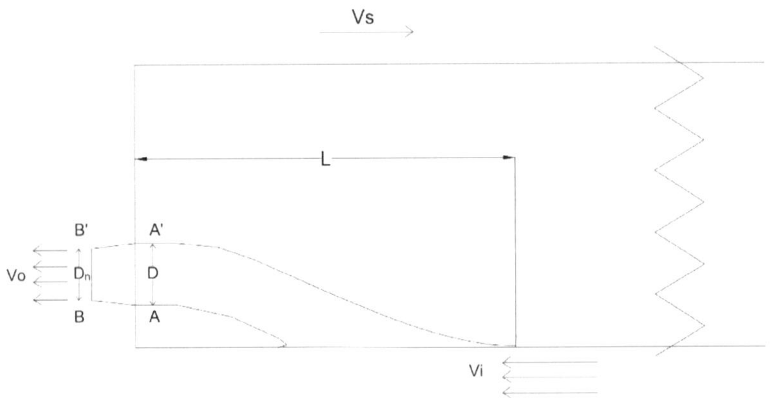

Figure 1. Parameters which govern the propulsive power (stern of the craft)

The effective power of the craft is considered equal to the propulsive power of the waterjet, i.e., $P_E = P_T$ as described in [20].

Then the diameter at the impeller plane of the waterjet, i.e. AA' in Figure 1, is computed using

$$D = 2\sqrt{\frac{P_T}{\pi V_S \varrho V_o V_S (V_o - V_i)}} \tag{3}$$

where,
D - Diameter at the impeller plane,i.e.(AA') [m]
ϱ - Density of the fluid [kg/m^3]
V_o - Outlet velocity [m/s]
V_i - Inlet velocity [m/s]
V_S - Design speed of the craft [m/s].

It is assumed the nozzle diameter (BB') is the same as the diameter of the impeller plane (AA'). When the diameter is known, the overall length L of the waterjet is determined as described in [21] using

$$L = 6.38D \tag{4}$$

Then, the two dimensional (2D) geometrical profile of the water jet is defined as described in [21]. The lengths L1-L10 as shown in Figure 2 are selected relative to the previously computed diameter of the impeller plane. Table 1 shows the selection of profile parameters.

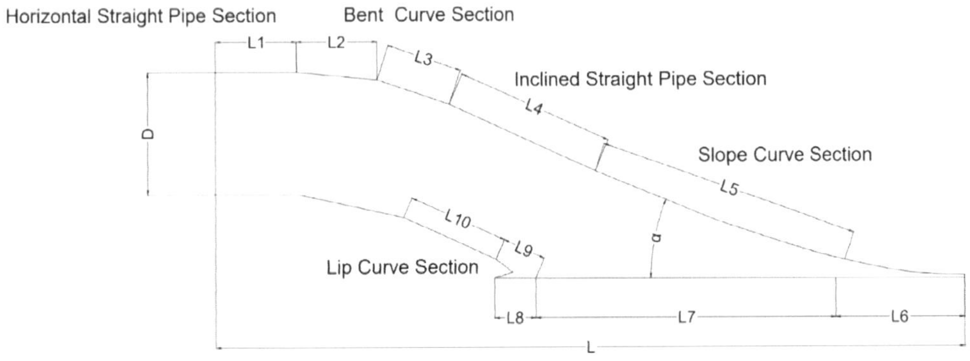

Figure 2. The 2-d geometrical profile of a waterjet

Table 1. Control Parameters of the 2-d geometrical profile of waterjet

Profile Parameter	Description	Value
L	Total Length	6.38 *D**
L1	horizontal straight-line section	0.68 *D*
L2	Bent curve section	0.68 *D*
L3	Bent curve section	0.68 *D*
L4	inclined straight-line section	1.36 *D*
L5	slope curve section	2.27 *D*
L6	slope curve section	1.14 *D*
L7	horizontal inlet	2.59 *D*
L8	lip curve section	0.36 *D*
L9	lip curve section	0.40 *D*
L10	inclined straight-line section	0.85 *D*

**D – diameter of the impeller plane*

The lengths L_1 to L_{10} are adjusted to appropriately fit into the stern shape of the hull while keeping the inflow dip angle (α) at 35°, the upper lip angle(β) at 4° as shown in Figure 3 and the Inlet Velocity Ratio (IVR) ($\frac{V_i}{V_s}$) between 0.67 to 0.87 as described in [21]. A tailor-made algorithm can be used to optimize the jet profile which is not in the scope of this work.

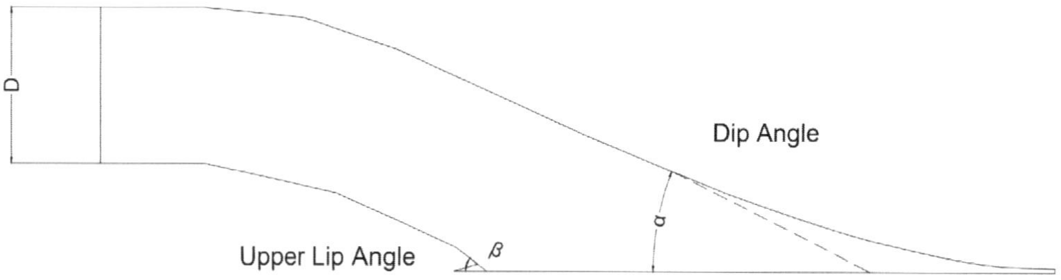

Figure 3. Definitions for the inflow dip angle (α) and the upper lip angle(β)

2.3 Scaling

After selecting the jet profile, both the hull and waterjet are scaled using Froude's law of comparison as described in [22, 23] using the same scale factor expressed as

$$\lambda = \frac{L_S}{L_M} \tag{5}$$

where,
L_S - Full-scale ship length [m]
L_M - Model-scale ship length [m].

The scale ratio is selected based on Table 2 compromising the model length of 1-2m as described in [24]. This model length range is selected to avoid too small models where scale effects and measuring accuracy occur and too large sizes which would be costly, complicated to manufacture, handle and where blockage effects also occur as described in [25].

Table 2. Scale ratios for different craft lengths

Length Range (m)	Scale Ratio
0 – 20	10
20 – 40	20

2.4 TESTING TANK

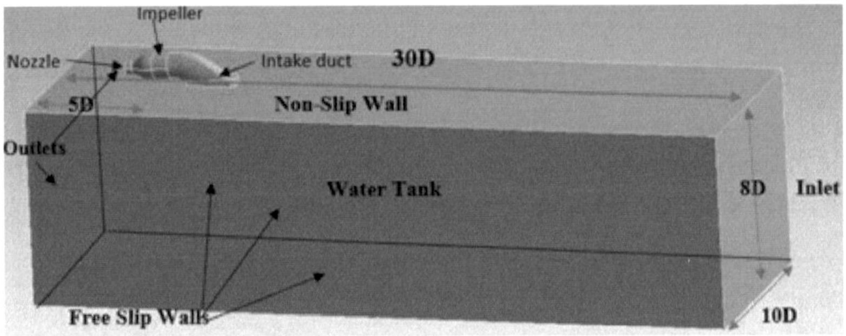

Figure 4. Tank size definition and refinement

The preliminary dimensions of the model testing tank are defined based on the diameter of the model waterjet calculated using Equation 3 and scaled down using Equation 5 as described in [26] (see Figure 4). Then the dimensions are further refined using a CFD-study to avoid wall, blockage and finite-depth-effects as described in [27]. Furthermore, the CFD study is simplified by using wall boundary conditions without using inlet and outlet boundary conditions of the water tank and fully submerging the waterjet. The waterjet is fully submerged to reduce the computational effort.

The thrust of the waterjet is used to refine the tank dimensions. It is defined as the force exerted by the fluid flow normal to the direction of the outflow. In the CFD study, the thrust of the waterjet is computed in an iterative process while altering the main dimensions of the tank, i.e. length, breadth and width, by a factor of 0.1 as follows

$$Length = D(1 + 0.1\lambda_\alpha) \cdot 30 \quad (6)$$

$$Width = D(1 + 0.1\lambda_\alpha) \cdot 10 \quad (7)$$

$$Breadth = D(1 + 0.1\lambda_\alpha) \cdot 8 \quad (8)$$

where,
$\lambda_\alpha = \pm 1, 2, 3, 4 \ldots$

Then the final dimensions of the tank are selected at the point where the thrust starts converging.

2.5 HULL-JET CONFIGURATION

Three hull-jet configurations are selected as depicted in Figure 5 In the first configuration, 5(a), an electric motor fixed inside a watertight box is coupled to the waterjet using a long propeller shaft. The effect of the hull is simulated in the second configuration, 5(b), however, the waterjet is placed outside the hull while coupling them with a propeller shaft secured inside a watertight tube. The third configuration, 5(c) resembles the actual jet installation.

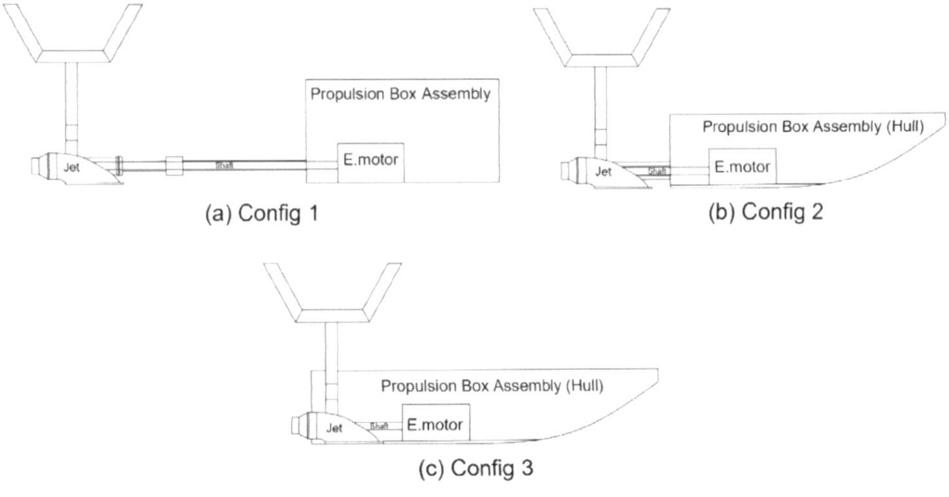

Figure 5. Configurations of hull-jet

An appropriate hull-jet configuration is selected based on the effect of the hull on the jet water intake, effect of the propeller shaft on the water flow and the mechanical elements for stability and vibration reduction.

A circulating pump loop setup placed vertically or horizontally as depicted in Figure 6 is used to vary the intake velocities to simulate the real case.

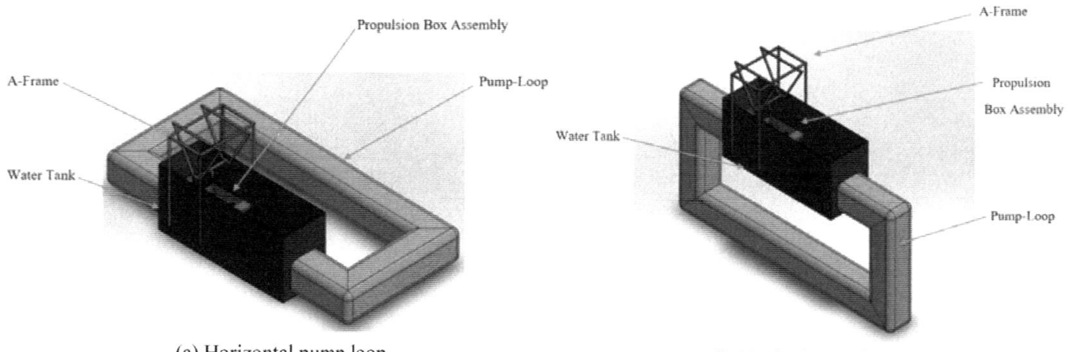

(a) Horizontal pump loop (b) Vertical pump loop

Figure 6. Circulation loop layouts

3. RESULTS

The model scale testing method was developed under four consecutive steps. i.e., hull, waterjet, scaling and testing tank. First, the resistance and power curves were computed for the design-speed curve of the selected SHSC. Then the diameter of the waterjet impeller plane was calculated considering the effective power is equal to the propulsive power of the waterjet. The profile of the waterjet was formed using the determined impeller plane diameter. Thereafter, an appropriate scale ratio was selected to compromise the model length between 1-2m. Then, both the hull and the water jet were scaled using the selected scale ratio. The preliminary dimensions of the testing tank were then defined based on the diameter of the waterjet impeller plane and further refined using a CFD study. Finally, an appropriate hull-jet configuration was selected considering the comparison given in Table 3.

Table 3. Comparison of hull-jet configurations

Parameter	Configuration 1	Configuration 2	Configuration 3
Effect of hull on the jet water intake	Comparatively low	Comparatively high	Substantially high, however, resembles real situation
Effect of the shaft on the water flow	Very high (Not the actual case)	Low (Not the actual case)	No (Actual case)
Mechanical elements for stability and vibration reduction	High complex	Moderate complex	Low complex

The third configuration, 5(c) was selected as the hull effect on the jet flow is realistic, there is no shaft effect on the flow and the disturbance from vibration is minimal.

The horizontal layout for the circulation loop is selected as cavitation testing is not a major consideration in this study.

4. A PROPOSAL TO EVALUATE WATERJET PERFORMANCE

In this study, a method is proposed to evaluate waterjet performance. Waterjet performance is presented at model scale as head, power and efficiency as shown in Figure 7.

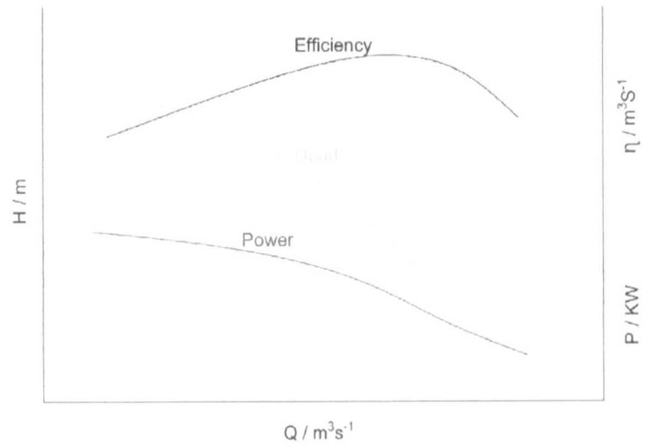

Figure 7. Performance output for a model-scale waterjet

The flow-rate range is selected using the speed range of the vessel which corresponds with the power range. i.e., the flowrate at the respective vessel's speed range and is calculated as

$$Q = AV_o \tag{9}$$

where,
Q - Flowrate [m^3/s]
A - Cross sectional area at the outlet calculated using D from Equation 4 [m^2]
V_o - Outlet velocity of the waterjet which corresponds to vessel-speed V_S [m/s].

The flow -rate range of the model-scale setup is calculated as described in [28] by

$$\frac{Q_s}{Q_m} = \lambda^{2.5} \tag{10}$$

where,
λ - Scale factor.

The performance output chart is plotted between the maximum flowrate, $(Q_m)_{Vmax}$ and minimum flowrate, $(Q_m)_{Vmin}$ corresponding to the operational range of the craft.

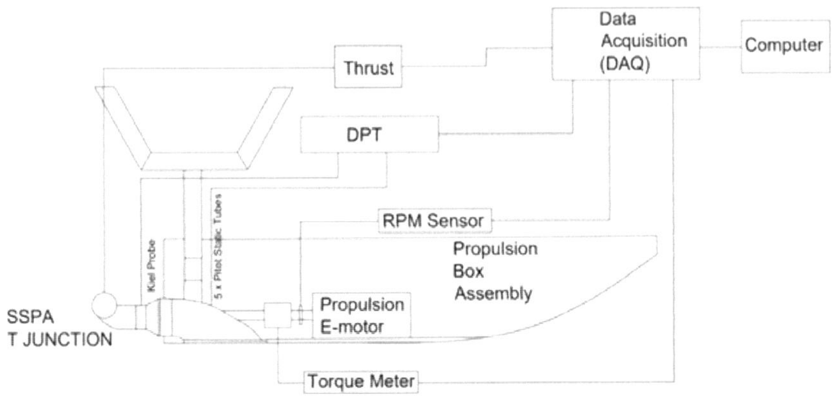

Figure 8. Final model-scale setup

The model scale setup is arranged as shown in Figure 8 and the parameters given in Table 4 are measured using respective methods.

Table 4. Measured parameter and methods

Parameter	Method
Pressure	Kiel Probes and Pitot Static Tubes
Velocity	Pitot-Static Tubes or Water Collection
Flowrate	DPT connected to the Kiel Probe
Head	DPT connected to a Pitot Static Tubes
Thrust	SSPA T-Junction Approach [29]

The thrust and torque are measured directly while head, power and efficiency are computed using pressure, velocity, flowrate and rpm as follows

a. Head is the difference of the pressure between two stations. For example, the head between station 6 and 1, i.e. (H_{61}) is computed as

$$Head = H_{61} = P^6 - P^1 \tag{11}$$

where,
P^6 *and* P^1 are the pressures at the respective stations.

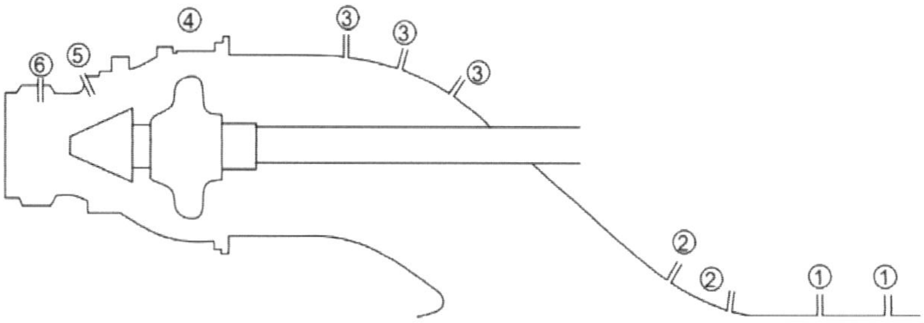

Figure 9. Pressures at different stations ($P^1 to\ P^6$) of the waterjet

b. Power of the waterjet propulsor is computed by

$$P = \tau\omega \tag{12}$$

where,
τ - Torque [Nm]
ω - Angular velocity [rad/s]

c. Pump efficiency is computed by

$$\eta_p = \frac{P_{PE}}{P_D} = \frac{\rho g Q_J H_{35}}{2\pi n Q_{shaft}} \tag{13}$$

where,
Q_J - Flowrate [m^3/s]
H_{35} - Head between stations 3 and 5 [m]
n - Revolutions per second, $N/60$

Q_{shaft} - Torque of the shaft [Nm]
The entire process of evaluating the performance of a waterjet using the model test set up is shown in Figure 10.

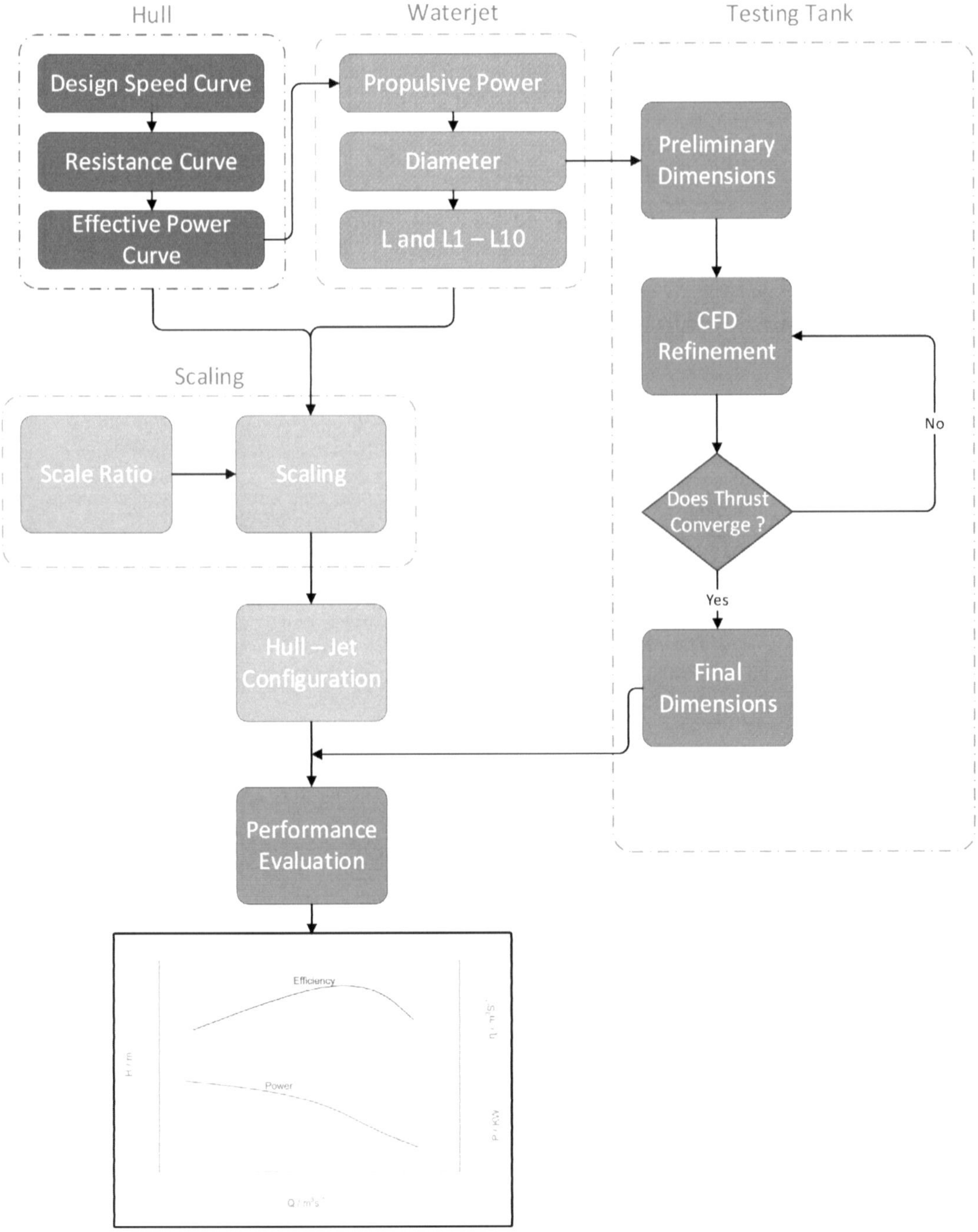

Figure 10. Waterjet performance evaluation process

5. DISCUSSIONS

A method was developed to evaluate performance of a waterjet using a model scale test set up. In the method, resistance and power curves of a selected SHSC was derived based on its design-speed curve. The diameter of the impeller plane of the waterjet was determined considering the effective power of the craft is equal to the propulsive power of the waterjet. Two-dimensional geometrical profile of the waterjet was then formed based on the impeller plane diameter. Then the hull and the waterjet were scaled using an appropriate scale ratio selected from the values recommended in the previous studies. The preliminary testing tank dimensions were then defined using the scaled impeller plane diameter and refined using a CFD study. Finally, a suitable hull-jet and circulation loop were selected. The work also proposed a procedure to evaluate the performance of a waterjet using the developed model scale test set up method.

In this study, Froude's law of comparison has been used to scale the hull and jet because both the inertia and gravity forces are predominant in free surface flows of the model scale setup. There are other scaling methods also available in this context, such as, Reynold's, Mach's and Weber's [25]. However, these methods cannot be complied in combination. Therefore, a scaling method must be selected based on the application and necessity. For this application, the selection of Froude's scaling method is advantageous as the introduced model scale setup being influenced by the formation of surface waves due to motions governed by gravitational forces.

It has been proposed in this work to first optimize the waterjet profile for the hull and then scale both using an appropriate scale ratio. However, when designing a craft an existing waterjet can be selected based on the design requirements and the hull can then be optimized for the waterjet profile. The proposed method is particularly advantageous for waterjet manufacturers since it facilitates designing high performance waterjets for existing proven high-speed hulls.

This method can also be used for other hull forms such as semi-displacement monohulls, multihulls with small waterplane area and hydrofoil craft with an appropriate method to compute resistance and effective power. Moreover, waterjet performance for multihull arrangements can be tested using a single demi-hull as described in [30].

The scale ratios proposed in this study is for the range of SHSC, i.e. length of the waterline below 24m and for a compromised model length between 1–2 m as described in [24]. It is to avoid too small models where scale effects and the accuracy are significant and too large models are expensive and complex to manufacture and handle and also prone to blockage effects when testing.

In this study for a SHSC, the preliminary dimensions of the tank are defined using the impeller diameter because the proposed method is most appropriate for a waterjet manufacturer. For other hull forms where the former method is not applicable, an alternative method is to define the preliminary dimensions of the tank based on the length of the craft as described in [31,32]. It is also possible to test and evaluate the waterjet using this model scale test set up even without a hull without considering the influence of the hull.

The most common layout is the vertical type for cavitation testing of propellers as described in [20] and the horizontal layout is preferred for testing of pumps and is described as a pump loop setup in [10] and tested using a test rig in [33].

The method is applicable for a wide range of hull-forms despite considering it for a planning hull-form. The method is currently in further development and could be used in any context at the preliminary design stage to evaluate performance. In the future, this method can be validated using existing waterjets and the water circulation loop can further be investigated.

6. REFERENCE

1. D. L. Blount and R. J. Bartee, "Design of Propulsion Systems for High-Speed Craft," *Marine Technology,* 1997.
2. M. Rudnicki and R. Sjogren, "Waterjet Propulsion for High-Speed Jets," *SAE Transactions,* 1974.
3. L. Arcand and C. Comolli, "Optimization for waterjet propulsion for High-Speed Ships," *Journal of Hydrodynamics,* 1968.
4. R. Svensson, S. Brizzolara and L. Grossi, "Trial result including wake measurements from the world's largest water installation," in *International conference on Waterjet Propulsion II, RINA*, Amsterdam, Netherlands., 1998.
5. C. Chesnakes, M. Donnelly, S. Jessup and F. Lanni, "Testing waterjet at range of scales," in *Proceedings of 27th Symposium on Naval Hydrodynamics*, Seol, South Korea, 2008.
6. M. Donnelly and S. Gowing, "Overview of recent development in testing of waterjets at NSWCCD," in *Proceedings of Waterjet Propulsion5*, London,UK, 2008.
7. J. Allision and C. Jiang, "Modern tools for waterjet pump Design and recent recent advances in the field.," in *International Conference on Waterjet propulsion II,RINA*, Amsterdam, Neterlands, 1998.
8. V. T. Terwisga, "Waterjet-Hull Interaction.," Delft University of Technology, Delft, Netherlands, 1996.
9. F. N., "Measurments of basic performances for waterjet propulsion systems.," *International Journal of Rotating Machinery 2(1),* 1995.
10. ITTC, "ITTC Reccomended Procedures and Guidelines,Testing and Extrapolation Methods High-Speed Marine Vehicles, Waterjets Waterjet system performance," in *ITTC*, 2005.
11. ITTC, ITTC – Recommended Procedures and Guidelines, Propulsive Performance Prediction. ITTC Report 7.5-02-05-03.1 (2011a).
12. ITTC, "Report of Specialist Committee on Validation of Waterjet Test Procedures," 1999.
13. ITTC, "Report of Specialist Committee on Validation of Waterjet Test Procedures," in ITTC Proc of 24th ITTC, 2005.
14. Bulten, Norbert. A Breakthrough in Waterjet Propulsion Systems. (2008).
15. Kandasamy, Manivannan & Georgiev, Svetlozar & Milanov, Evgeni & Stern, Frederick. Numerical and experimental evaluation of waterjet propelled delft catamarans. (2011).
16. O. Faltinsen, Hydrodynamics of High-Speed Marine Vessels, Hydrodynamics of Marine Vehicles, Cambridge: Cambridge University press, 2006.
17. O. Faltinsen, "Hydrodynamic features of High-Speed Vessels," in *Hydrodynamic features of High-Speed Vessels*, Ships and Offshore Structures, 2006.
18. D. Savitsky, Hydrodynamic Design of Planing Hulls, The Society of Naval Architects and Marine Engineers, 1964.
19. E. V. Lewis, Principles of Naval Architecture, vol. II, 1988.
20. J. Carlton, Marine Propellers and Propulsion., 2018.
21. J. Weixuan , L. Cheng, D. Zhang, B. Zhang, Y. Su and C. Wang, "Optimal Design of Inlet Passage for Waterjet Propulsion System Based on Flow and Geometric Parameters," *Advances in Materials Science and Engineering,* 2019.
22. W. Froude, R. Gawn and A. Duckworth, "The papers of William Froude,1810-1879. With a Memoir by Sir Westcott Abell and an Evaluation of William Froude's work by EWL Gawn," Institution of Naval Architects, 1955.
23. W. Froude, "Comparative resistances of long ships of several types.," *Transactions from Institute of Naval Architecture,* 1876.
24. ITTC, "Recommended 7.5-01-01-01 procedures and guidelines- ship model tests," ITTC, 2011.
25. E. Huse, "Experimental Methods in Marine Hydrodynamics," NTNU, Trondhiem, Norway, 1999.
26. R. Y. Huang, Y. Dai, X. Luo, Y. Wang, T. Du and C. Huang, "Investigations ito the unsteady internal flow characteristics for a waterjet propulsion system at different cruising speeds," *Ocean Engineering,* 2020.
27. ITTC, "Reccomended procedures and guidelines-Resistance Test," ITTC, 2011.
28. L. Castillo and J. Carrillo, "Analysis of the Scale Ratio in Nappe Flow Case by Means of CFD Numerical Simulation.," 2013.
29. L. Da-Qing and M. Brown, "Advanced tools for waterjet performnce prediction," in *SSPA*, 2015.
30. J. Rovere, "Catamaran Resistance from Tests on a Single Demi-hull," in *International Conference on Fast Sea Transportation*, Australia, 1997.
31. ITTC, "Practical Guidelines for Ship Resistance CFD," ITTC, 2014.
32. S. Kewei , G. Chunyu, S. Cong, W. Chao, G. Jie, L. Ping and W. Lianzhou, "Simulation strategy of the full-scale ship resistance and propulsion performance," *Engineering Applications pf Computational Fluid Mechanics,* 2021.
33. Z. Pan and Y. Ni, "Scaling of on-board characteristics of waterjet propulsion based on bench thrust testing data and loop test results," *Advances in Mechanical Engineering,* 2021.

7. ACKNOWLEDGEMENTS

The authors are grateful to Dr. Pahansen de Alwis of SSPA Sweden AB/Research Institute of Sweden (RISE)/KTH Royal Institute of Technology Sweden for encouraging to write this paper and for his invaluable advice and guidance without which this paper would not have been possible. Also special thanks to Dr. P.A.B.A.R. Perera and Dr. R.A.C.P Ranasinghe of University of Moratuwa, Sri Lanka, Dr. Nirman Jayaratne and Dr. Apsara Abeysiriwardhane of Australian Maritime College, University of Tasmania for their advice and suggestions. Lastly, we would also like to thank Mr. Dimantha Harshappriya of NTNU, Norway and Mr. Rohan Gamini Godhahewa Lecturer (Boat Building Technology) at Ocean University of Sri Lanka.

8. AUTHORS BIOGRAPHY

KGDL Dias holds the current position of Design Engineer at Walkers Shipyards. He is responsible for handling New Building Projects, preparing initial designs and calculations for newbuilding, preparing technical specifications, design reports and reviews, cost estimation, preparing general engineering drawings and docking arrangements. His previous experience includes working as a trainee engineer at Colombo Dockyard PLC and Sri Lanka Ports Authority.

TS Ruwandi has recently graduated from Ocean University of Sri Lanka. She is interested in hydrodynamic research on propulsion systems and high-speed craft. Her previous experience includes working as a trainee engineer at Colombo Dockyard PLC and Sri Lanka Ports Authority.

SASBS Suraweera holds the current position of Quality Control Engineer at Walkers Shipyards. He is responsible for quality control and assurance of ship building, ship repair and general engineering works. His previous experience includes working as a trainee engineer at Colombo Dockyard PLC and Sri Lanka Ports Authority.

MODIFICATION OF MOMENTUM DISTRIBUTION WITHIN THE THREE NOZZLES OF THE WATERJET AZIMUTH PROPULSION SYSTEM

R Zhang, P Yan and **J Zhou**, Marine Design & Research Institute of China, China

SUMMARY

This paper focuses on the thrust direction misalignment and lateral force caused by the rotational flow at the impeller outlet of the waterjet azimuth propulsion system. Momentum distribution within the three nozzles is found to be the direct reason. Two methods, i.e. area match between the left nozzle and right nozzle and adding guide vanes inside the middle nozzle are proposed and their effects are analysed respectively based on numerical simulation. Results show that an appropriate area ratio can eliminate thrust direction misalignment with a small increase in lateral force. Guide vanes can effectively reduce lateral force by 56% without affecting aligned thrust direction.

1. INTRODUCTION

As a special type of waterjet propulsion, the waterjet azimuth propulsion system can provide equal thrust on the direction over 360° and do not add any appendage to the ship to change its drag characteristic. As a result, it is perfectly suitable for shallow water condition and high manoeuvrability request. Up to now, the waterjet azimuth propulsion system has been used in various military crafts either as a main or an auxiliary prolusion unit. And several types of product series have been developed by different manufacturers.

Odetti collected and compared products from different manufacturers [1]. Although these systems share a similar working principle, their structures and arrangements differ significantly. In comparison, Schottel pump-jet (SPJ) demonstrates least space occupation. Its pump shaft is coaxial with the azimuth shaft, thus reducing bulkiness [1]. So Odetti designed his own Pump-Jet Module (PJM) based on SPJ.

Main components of SPJ include the grid, impeller, diffusion chamber and three nozzles. The diffuser chamber plays the role of transforming water kinetic energy to water pressure energy and forming high speed waterjet at the nozzles. In early SPJ products (T series), guide vanes after the impeller were designed to eliminate the rotational flow at the impeller outlet [2]. While in recent SPJ products (D series), guide vanes are abandoned and the so-called "ring diffuser" is used instead. They claimed that such change increases flow areas, reduces flow velocity and noise and increases thrust by 12%-15% approximately [3][4]. Odetti's PJM has similar design but adds three guide vanes inside the middle nozzle to deal with the rotational flow derived from the impeller rotation. However, these guide vanes' effects were not discussed in his paper [1]. Liu also designed his waterjet azimuth propulsion system based on SPJ and paid much attention to impeller parameters rather than the diffusion chamber [5].

In summary, although there have been various products of the waterjet azimuth propulsion system, very few studies exist in the literature on pump-jet design and functioning, not to mention modification of the diffusion chamber. But actually, the rotational flow at the impeller outlet results in uneven momentum distribution within the three nozzles. Such uneven distribution leads to the appearance of lateral force and thrust direction misalignment, thus weakening the manoeuvrability of the waterjet azimuth propulsion system.

In this paper, two methods, i.e. area match between the left nozzle and right nozzle and adding guide vanes inside the middle nozzle, are proposed and their effects on lateral force and thrust direction misalignment are investigated based on numerical calculation.

2. MODEL AND METHOD

The research model is established as shown in Figure 1, including the grid, impeller and diffuser chamber. The three nozzles are integrated with the diffuser chamber. The impeller inlet diameter is D1=300 mm and the flow passage diameter is D=650 mm.

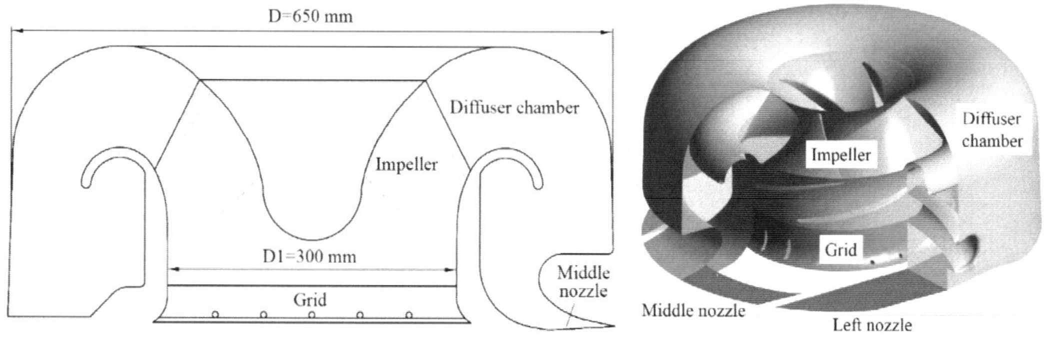

Figure 1. The research model

The basic diffuser chamber model does not have any guide vanes and areas of the two-side nozzles are also the same. This model will be taken as the reference model. Obviously, affected by the rotational flow at the impeller outlet, waterjet of the reference model will deflect to the direction of impeller rotation, as shown in Figure 2. This will lead to lateral force. In addition, the flow rate of the nozzle whose waterjet direction is the same as the rotational flow will be larger than that is opposite. This will lead to thrust direction misalignment, i.e. thrust direction deviating from the centre line. For easier understanding on following flow analysis, the coordinate system shown in Figure 2 is established.

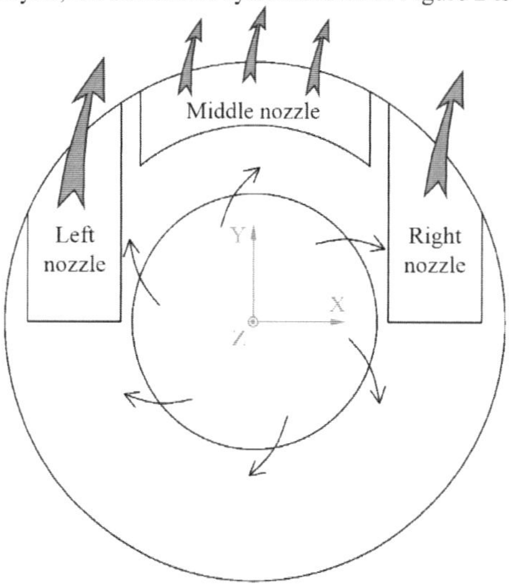

Figure 2. The schematic diagram of nozzle waterjet flow

Area match (between the two-side nozzles) is applied to eliminate thrust direction misalignment. Different area ratios from 0.85~1 between the left nozzle and right nozzle (total area keeps the same) are tested. The one with minimum thrust direction misalignment will be selected as model A. And then guide vane will be added to model A to reduce lateral force and forms model B. Guide vane's effects on lateral force reduction will be analysed. The variation of models is shown in Table 1.

Table 1. Modification of the diffuser chamber

Model	**Area ratio**	**Guide vane**
Reference	1	No
Test of A	0.95	No
	0.9	No
	0.85	No
A	Best area ratio	No

B	Best area ratio	Yes

Modification of the diffuser chamber is based on numerical calculation. The computational domain is shown in Figure 3, including the research model and water domain. Mesh construction is completed in ANSYS ICEM using tetrahedral grid for the internal domain and triangular prim grid for wall surfaces. Calculation is performed in ANSYS CFX and important setup information is shown in Table 2. Considering the greatest advantage of the waterjet azimuth propulsion system is its high manoeuvrability rather than working as a main thruster [1], the mooring condition (with zero freestream velocity) is selected as working condition in this paper.

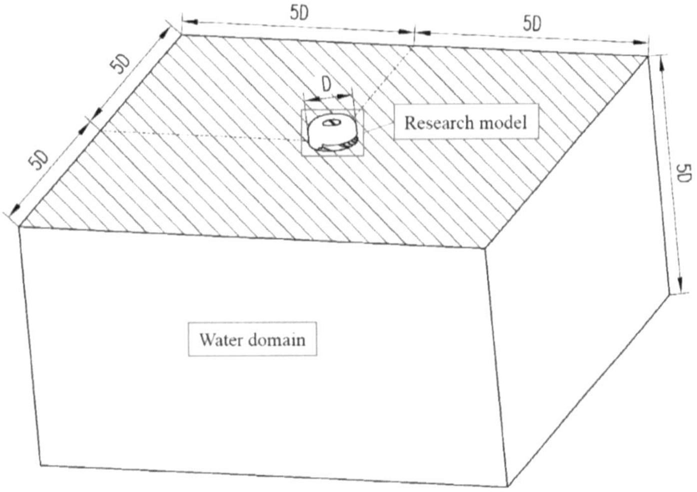

Figure 3. The computational domain

Table 2. Setup information in ANSYS CFX

Item	Description
Rotation speed of the impeller	1700 r/min
Freestream velocity	0
Material	Water (25 ℃)
Turbulence model	SST k-ω
Advection scheme	High resolution

3. RESULTS AND DESCUSSION

3.1 AREA MATCH

The focus of this paper is to improve the thrust characteristic of the waterjet azimuth propulsion system. Formula (1) is usually used to calculate the thrust of traditional waterjet propulsion system [6].

$$T = \rho Q(V_J - \alpha V_0) \quad (1)$$

In formula (1), T is the thrust, ρ is the material density and Q is the flow rate of propulsion system. V_J and V_0 are waterjet velocity and ship velocity, i.e. freestream velocity, respectively. α is the inflow influence factor. For the waterjet azimuth propulsion system under mooring condition, flow is sucked in almost vertically, which does not contribute to the thrust. Therefore, αV_0 can be considered as 0. In addition, the nozzle outlet is almost horizontal, and water is ejected obliquely from the nozzle. So there will be an angle β between V_J and V_T. V_T is the velocity component in thrust direction. Meanwhile, Q can be expressed as AV_V, in which A is the nozzle outlet area and V_V is the vertical velocity component. As a result, formula (1) can be expressed as below for the waterjet azimuth propulsion system:

$$T=\rho QV_J\cos\beta=\rho AV_V V_T \quad (2)$$

As shown in Figure 2 and described above, the rotational flow at the impeller outlet will make the flow rate and thus velocity in the left nozzle larger than that in the right nozzle. According to formula (2), the thrust of the left nozzle will exceed that of the right nozzle. This will further lead to thrust direction misalignment, i.e. the direction of total thrust deviating from the centre line. The thrust deviation distance, x_F, is used to quantitatively describe thrust direction misalignment, which is expressed by formula (3).

$$x_F=\frac{\int \rho x\cdot V_T dQ}{\int \rho V_T dQ} \quad (3)$$

According to the cause of thrust direction misalignment, area ratio between the left and right nozzle is a key influence factor. Based on numerical calculation, the relationship between the area ratio and thrust deviation distance, x_F, is shown in Figure 4. In Figure 4, variation of the normalized lateral force (by the thrust), F_S/T, is also plotted as area ratio changes.

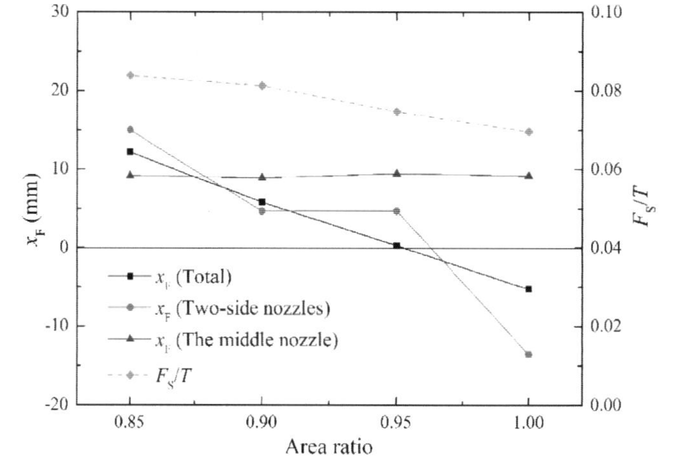

Figure 4. Variation of thrust deviation distance and normalized lateral force as area ratio

The area ratio of the reference model is 1 and there is a significant thrust direction misalignment, i.e. total x_F reaches approximately -5.24 mm (the minus sign indicates the negative X-axis shown in Figure 2). This is mainly caused by the larger flow rate of the left nozzle than the right nozzle, as show in Figure 5. As the area ratio decreases, the flow rate of the middle nozzle keeps almost constant, corresponding to the almost unchanged x_F of the middle nozzle in Figure 4. Meanwhile, the flow rate of the left nozzle decreases and that of the right nozzle increases. Such redistribution of flow rate, i.e. momentum, makes x_F of two-side nozzles increases. The increased x_F of two-side nozzles and the almost constant x_F of the middle nozzle contribute to the increase of total x_F. The point of area ratio with zero x_F, i.e. the area ratio of 0.95, is required by model A, indicating the elimination of thrust direction misalignment.

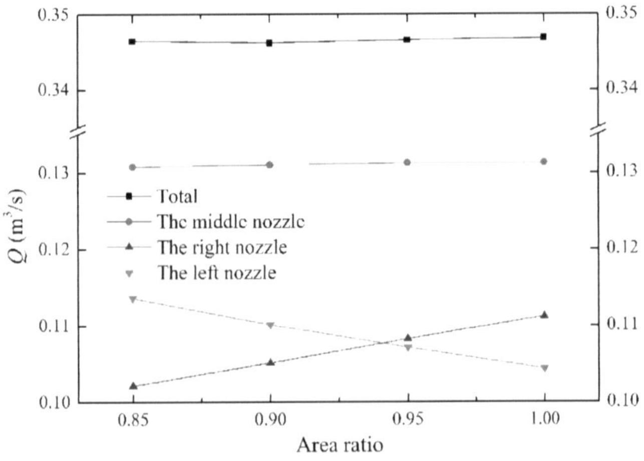

Figure 5. Variation of the flow rate, Q, as area ratio

Derived from formula (2), the thrust intensity, $I_F = \rho V_V V_T$, is defined to describe the distribution of thrust over the nozzle outlet. High I_F means high thrust. As shown in Figure 6, affected by the rotation flow in Figure 2, the two high I_F regions in the middle nozzle deflect to the right, corresponding to the positive x_F in Figure 4. The high I_F region in the left nozzle deflects to the left, which is the reason for the negative x_F in two side nozzles and total x_F. As the area ratio decreases, this high I_F region moves to the right, making x_F increase and thrust direction misalignment eliminated.

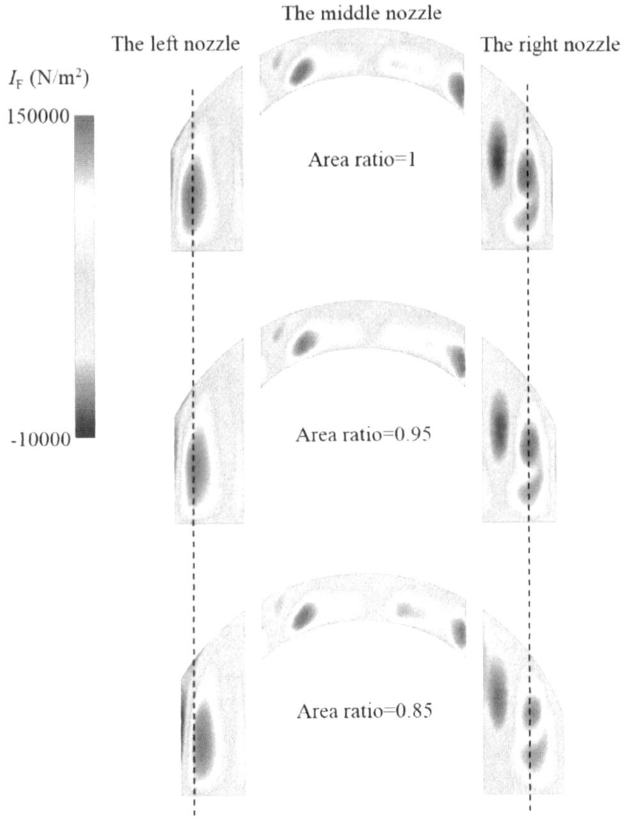

Figure 6. The thrust intensity distribution under different area ratios

3.2 GUIDE VANE

From Figure 4, although decreasing area ratio shows significant effects on thrust direction misalignment elimination, it will also cause additional lateral force. For model A, F_S/T reaches 0.075. From Table 3, it is found that the middle nozzle contributes the most to lateral force. So guide vanes are added inside the middle nozzle to decrease lateral force, as shown in Figure 7.

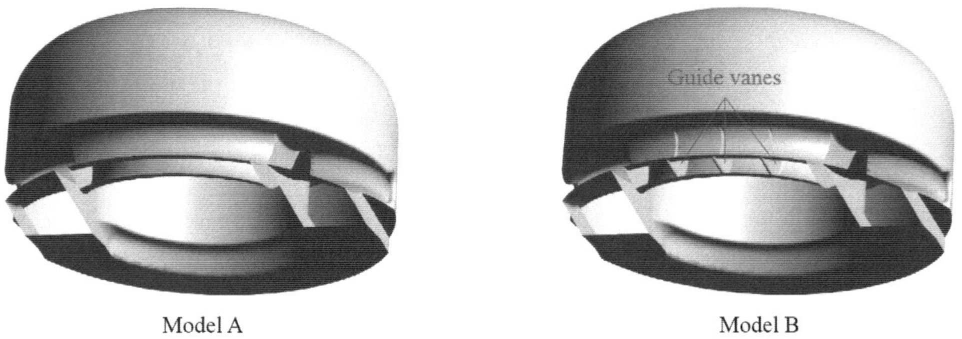

Figure 7 Model B with guide vanes

Table 3. Force characteristic comparison between model A and model B

	F_S/T	Contribution from different nozzles		x_F (mm)
		The middle nozzle	Two-side nozzles	
Model A	0.075	82%	18%	0.30
Model B	0.033	79%	21%	0.35

Compared with model A, there is a significant drop of 56% on F_S/T for model B with guide vanes. Contribution from different nozzles and x_F keep almost unchanged. This is because guide vanes change flow direction in the middle nozzle with little effects on flow rate and momentum distribution. Figure 8 shows guide vanes' effects on jet streamlines.

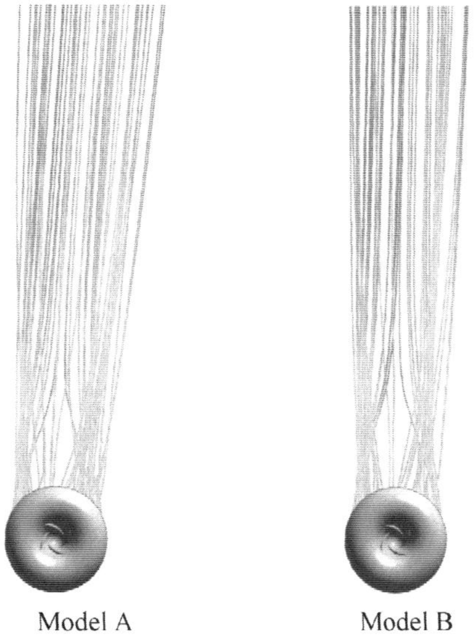

Figure 8. Comparison of streamlines between model A and model B

4. CONCLUSIONS

Momentum distribution mechanism within the three nozzles of the waterjet azimuth propulsion system is investigated in this paper based on numerical calculation. Results show that affected by the rotational flow at the impeller outlet, waterjet of the nozzles will deflect to the direction of impeller rotation. This will lead to lateral force. In addition, the flow rate of the nozzle whose waterjet direction is the same as the rotational flow direction will be larger than that with opposite direction. This will lead to thrust direction misalignment, i.e. thrust direction deviating from the centre line. According to this, two methods, i.e. area match between the left nozzle and right nozzle and adding guide vanes inside the middle nozzle, are proposed to deal with lateral force and thrust direction misalignment. It is found that an appropriate area ratio can eliminate thrust direction misalignment with a small increase in lateral force. Guide vanes can effectively reduce lateral force by 56% without affecting aligned thrust direction.

5. REFERENCES

1. Angelo Odetti, Marco Altosole, Gabriele Bruzzone, Massimo Caccia, Michele Viviani. 'Design and Construction of a Modular Pump-Jet Thruster for Autonomous Surface Vehicle Operations in Extremely Shallow Water'. *Journal of Marine Science and Engineering*, 2019.

2. SCHOTTEL GmbH. 'Schottel Product Guide', *SCHOTTEL GmbH & Co. KG*, 1999

3. SCHOTTEL GmbH. 'Schottel Product Guide', *SCHOTTEL GmbH & Co. KG*, 2021

4. SCHOTTEL GmbH. 'Schottel Report', *SCHOTTEL GmbH & Co. KG*, 1999

5. Wenhao Liu. 'Plate type rotary jet propulsion – the design and performance study', *Jiangsu University of Science and Technology*, 2016

6. Lixiang Wang, Youlin Cai. 'Design Theory and Technology of Waterjet Propulsion and Propulsion Pump', *Shanghai Jiao Tong University Press*, 2018

6. AUTHORS BIOGRAPHY

Ruizhi Zhang holds the current position of assistant engineer at Marine Design & Research Institute of China. He is responsible for the research and development of waterjet propulsion system. His previous experience includes the design and development of axial-flow type waterjet propulsion units, etc.

Yan Pneg holds the current position of senior engineer at Marine Design & Research Institute of China. He is responsible for the research and development of waterjet propulsion system. His previous experience includes the design and development of axial-flow and mixed-flow type waterjet propulsion units, etc.

Jiajian Zhou holds the current position of researcher at Marine Design & Research Institute of China. He is responsible for the research and development of waterjet propulsion system. His previous experience includes the design and development of axial-flow and mixed-flow type waterjet propulsion units, etc.

WATERJET IN BOW: PROPULSION AND GENERATION

G García Lage, University of A Coruña, Spain

SUMMARY

In this article, a general approach is made to the positive impact that the recovery of kinetic energy by means of bow thrusters in precautionary maneuvers and approach with high inertia and in the presence of wind has on the carbon footprint of a ship. With the pre-design of this energy collector, the power/speed and thrust/speed graphs are generated for a fluid vein of seawater, obtaining values with them of reduction of consumption and CO2. With the beginning of the first hydrodynamic tests on the CFD in a pipe, a basic prototype of this jet propeller is configured, like a pumped/turbined reversible hydroelectric plant, with the aim of continuing in the future different maneuvers that allow testing the hull of the ship in the presence of wind seeking to increase energy efficiency.

NOMENCLATURE

A	Area	*t*	Tons (Tm)
CO2	Carbon dioxide	*L*	Length (m)
Cp= K	Coefficient	MDO	Medium Diesel Oil
CFD	Computational Fluid Dinamyc	*MW*	Megawatts
m^3/s	Cubic meter per second	W_A	Maximum power
ρ	Density of water (kg m-3)	T_A	Maximum torque
Ø	Diameter (m)	*m*	Meter
E	Energy	*m/s*	Meters per second
Ccomb	Fuel consumption	*P*	Power (*kW/h*)
IFO380	Intermediate Fuel Oil	*R*	Radio (m)
IMO	International Maritime Organization	V^3	Speed cube
KERS	Kinetic Energy Recovery System	V^2	Speed squared
kg	Kilograms	*V*	Speed
kW	Kilowatts	m^2	Squared meter
kW/h	Kilowatts/hour	*Π*	PI (3,141516)

1. EFFICIENCY AND DISRUPTIVE INNOVATION.

Currently, in general, one of the priorities to be resolved at a global level is the high dependence on fossil fuels due to their high environmental impact, geostrategic instability, and scarcity. In a globalized economic system, dominated by Economies of Scale, maritime transport plays a key role in the distribution of goods with a great impact on International Trade, emitting large amounts of CO2.

Given this scenario, the IMO, among other institutions such as the European Union, have established a series of laws, regulations, and criteria in maritime transport to reduce these impacts on the environment or in urban port areas with the aim of promoting alternative systems based on renewable energy, autonomous systems, efficiency, etc. The Next Generation plans, for example, are a key indicator to understand the process of structural change that will take place in the coming years in advanced societies following the sustainable development goals set by the UN and where the disruptive innovation of concepts, alternative systems or processes are key to achieving the model change.

In this energy, technological, social, and economic transformation, new opportunities appear to establish new frameworks in the design of ships until today not considered in the search to increase global efficiency in maritime transport, favoring a multitude of synergies on board with the aim of reduce the dependence on fossil fuels.

Therefore, this research presents a planned route of a ship and a first bow propulsion prototype, with or without water jet, integrated in its new location in a standard bow to carry out CFD computational calculation tests, refining the hydrodynamic design of the different components trying to reach the generation-propulsion limit values after considering different scenarios, or problems, that may arise in the future development of this doctoral research.

Currently, there is a wide range of bow thrusters on the market with a multitude of designs, advantages, and disadvantages, where they adapt to the operating and performance criteria established for each ship: fixed tubed bow thrusters, retractable bow thrusters, water jet located on the side (Omni thruster), etc. These engines leave the factory designed with high efficiencies to fulfill a single mission in the bow: to propel.

At the same time, in the State of the Art, designers or companies such as Hyundai Heavy Industries [1] have patented a series of specific technologies for ships that replicate the operation of a conventional hydroelectric turbine by having internal ducts in the hull, or outside of it, channeling a flow of seawater towards the turbine generating electricity when sailing and, in many cases, aided by sail systems but under these configurations they are forced to maintain conventional bow thrusters, assuming twice the installation cost when coupling both systems separately.

After analyzing different electrical power generation systems with marine turbines used with the advance of the ship in the presence of wind, the development of an intermediate alternative began, that is, taking advantage of the bow thrusters themselves, which have large diameters, to carry out a secondary power generation function. Like a pumped/turbined reversible hydroelectric plant where the best possible performance ratio will have to be sought, recovering a part of the kinetic energy transmitted to the water by the ship in its advance. There are other studies of interest [2] where pumps are placed in holes that cross the hull, in strategic places, with the aim of pushing the water finding the same velocity ratio in the fluid layer, inside and outside the hull, greatly reducing the amount of energy required by the ship to push large amounts of water with its progress.

2. ENERGY RECOVERY SYSTEM IN THE SEA.

Energy from the sea has great potential but its use is still incipient, although legislative measures are being taken in different coastal countries for a better implementation in those locations where there is profitability. The energy supplied by the sea has better wind work coefficients and this is indicated by the professor Fernández Diez [3].

Energy density

-wind: 15m/s → 2kW/m2
-ocean current: 3m/s →14kW/m2.

If the regimes at low speed of the marine currents (2 to 3m/s) are exchanged for the maximum speeds (5 to 24m/s) generated by the ship with its progress, an approximate value of energy recovered in navigation is obtained. In the worst design and working condition of all. That is, a free propeller, without a tube, that generates tidal energy installed under the keel of a large ship that sails at cruising speed and reproduces a maneuver approaching a coast. It takes approximately 1 or 2 hours after starting the main engine speed reduction maneuver until the complete stop in port. Being the inertia of the ship in its advance a determining factor in the entire approach process.
For all these reasons, with a propeller design inside a pipe, its integration into the ship's hull, with a tailwind in transoceanic navigation or speed reductions where inertia plays an important role and taking into account a multitude of parameters, no all can be considered in this thesis due to the material and temporary impossibility, a better performance can be obtained by searching for the equivalent hours of operation of the device for different ships according to their dimensions and operational destination without giving up the propulsive capacity.

In tidal turbines applying the following formulation, the power values of the fluid vein of a marine current that can be extracted by the turbine are obtained as shown below. According to reference [4] it proceeds as follows.

- TSR; Tip Speed Ratio is the relationship between the speed of the blade tip and the speed of the fluid, for horizontal axis turbines the value of TSR ranges between 4 and 5.

$$TSR = \frac{\text{shovel speed}}{\text{fluid velocity}} \tag{2}$$

- Coefficient Cp, percentage of energy removable by a turbine of a fluid over the theoretical maximum, a value of Cp = 0.3 is granted, with 30% of the energy being usable. Although in high performance turbines this coefficient can reach the values of Cp = 0.4-0.5. (In duct).

- Power extracted, is calculated through the following expression.

$$P = ½.\ Cp.\ \rho.\ A.\ V^3 = kW/h \tag{3}$$

Where:
Cp, coefficient 0.3

ρ, sea water density, 1025 (kg·s/m^4)
A, area occupied by the rotor
V^3, cube speed.

- Rotation speed rotor

$$Wr = \frac{\text{Vrotor}}{\text{Rrotor}} = \frac{rad}{seg} = rpm \quad (4)$$

- Torque

$$T = \frac{\text{P}}{\omega} = \frac{W}{rad/sg} = N.m \quad (5)$$

2.1 (a) Example of steps to follow in pre-design of a Marine Turbine.

- Area calculation

$$A = \frac{\pi.D^2}{4} = m^2 \quad (6)$$

- Max Push

$$T_M = \left(\frac{\rho V^2}{2}\right) A \quad (7)$$

- Apply coefficient Cp (K) and obtain the net thrust

$$T_A = K\,.T_M \quad (8)$$

- Power that the turbine can extract from the fluid vein, in HP.

$$W_M = \frac{\frac{1}{2}\rho V^2\left(\frac{\pi D^2}{4}\right)V}{75} \quad (9)$$

- Therefore, to obtain the net power and in kilowatts we do the following Coefficient K= **0,3.**

$$W_A = K.W_M.0{,}735 \quad (10)$$

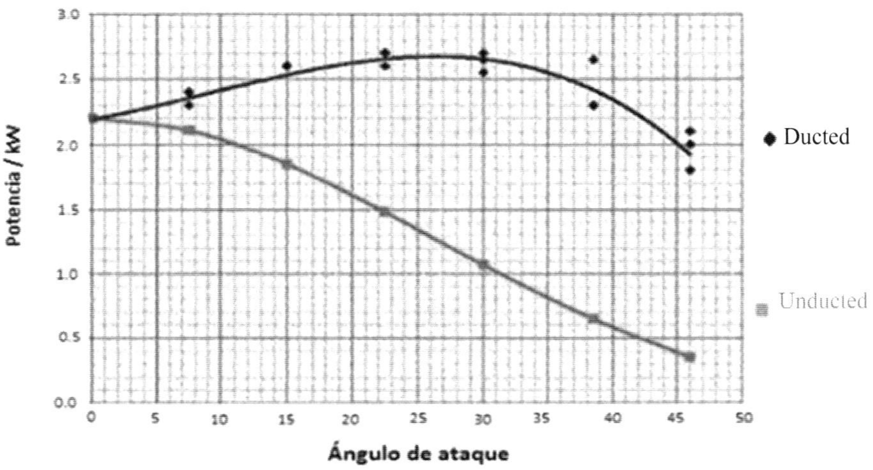

Figure 1. Performance Power with or without pipe.
Source: Glasgow Tank tests Mayo 2004. Comparison of Ducted & Unducted Turbines Angled to the flow.
Cp(K)= 0,35

2.1 (b) Experimental Evaluation: Power.

The calculations obtained by applying reference [4] are summarized in the following graphs. Graphical representation Power Applying the calculation method for a fluid vein of seawater with speeds from 24 knots to 4 knots and propeller diameters of 4*m* and 2*m*, is obtained for a coefficient *Cp (k)* = 0.3.

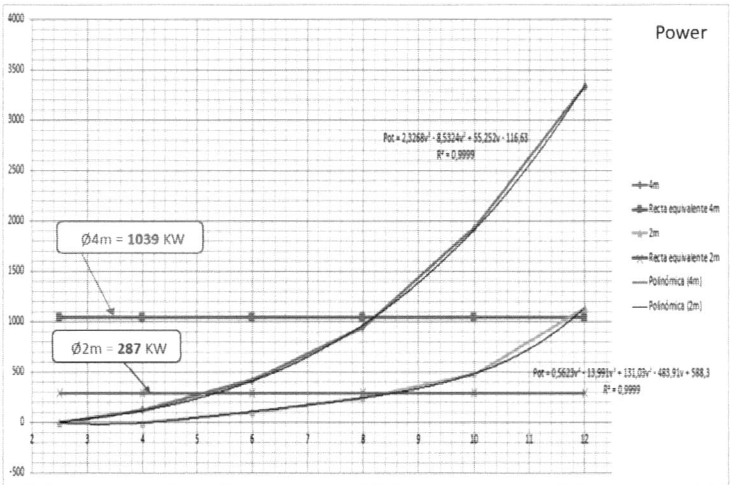

Graph 1. Representation of the average net power by means of curves and equivalent area lines for speeds of 12*m/s* at 2.5*m/s* and diameters of 4 and 2 *m*.

				v	pot
	Función velocidad	12-0,0017t			
				12	1039
	Función potencia	2,3268v^3-8,5324v^2+55,252v-116,63		10	1039
4m				8	1039
	Integrando entre 2,5 y 12	9874		6	1039
				4	1039
	Resolviendo para una recta de área equivalente		1039kW	2,5	1039
	Función velocidad	12-0,0017t		v	pot
				12	287
	Función potencia	0,5623^4-13,991v^3+131,03v^2-483,91v+588,3		10	287
2m				8	287
	Integrando entre 2,5 y 12	2723		6	287
				4	287
	Resolviendo para una recta de área equivalente		287kW	2,5	287

Table 1. Calculation of the net power for a line of equivalent area.

2.1 (c) Graphical representation: Push.

Applying the calculation method for a fluid seawater vein with speeds from 24 knots to 4 knots and propeller diameters of 4m and 2m, it is obtained for a coefficient *Cp (k)* = 0.3.

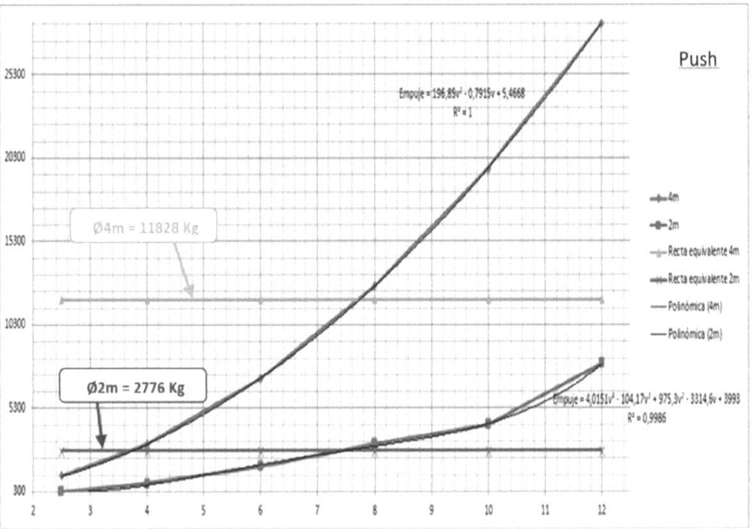

Graph 2. Representation of the average net thrust by means of curves and equivalent area lines for speeds of 12*m/s* at 2.5*m/s* and diameters of 4 and 2 *m*.

				v	Empuje
	Función velocidad	12-0,0017t			
				12	11828
	Función empuje	116,85v^2-0,7915v+5,4668		10	11828
4m				8	11828
	Integrando entre 2,5 y 12	112362		6	11828
				4	11828
	Resolviendo para una recta de área equivalente		11828kg	2,5	11828
	Función velocidad	12-0,0017t		v	Empuje
				12	2776
	Función empuje	4,0151^4-104,17v^3+975,3v^2-3314,6v+3993		10	2776
2m				8	2776
	Integrando entre 2,5 y 12	26371		6	2776
				4	2776
	Resolviendo para una recta de área equivalente		2776kg	2,5	2776

Table 2. Calculation of the net power for a line of equivalent area.

Note: All calculations have been based on the worst design conditions, it is possible to obtain better results of electrical production under optimal design conditions. The study is based on a pipe less propeller analyzing the least favorable working condition of all, penalizing the *Cp* coefficient (formula 3 of this work).

It is key, therefore, to keep the red line (Graph 1) as high as possible on the graph, seeking the best possible relationship of factors in its design when integrating it into the ship, maximizing the time that the system can be activated. Where the presence of a tailwind with the ship at high speed, among other maneuvers to study, are relevant to obtain high values of energy recovered from the sea and returned to the ship. This research presents a reduced analysis of the approach to the coast in the worst working conditions that allows establishing starting criteria to propose different hypotheses of scientific interest in its development.

3. SUSTAINABILITY DURING NAVIGATION.

3.1 INERTIA

When sailing a boat, one of the most demanding manoeuvres is to approach the shore and come to a complete stop. It is essential to carry out actions to correct deviations in the ship and manage the hours invested in manoeuvres from another operational vision, more sustainable, replicating the operation of a renewable electricity generation plant to reduce consumption and pollution through a similar process to KERS of the transport industry.

The IMO (International Maritime Organization) includes in its regulations [5] the minimum design distance necessary to stop a ship with reverse gear at full speed and it must not exceed 15 lengths, except in special large ships that can extend this distance. That is to say, the forces present in this braking manoeuvre with reverse gear at full speed implies a minimum of 15 lengths of a large ship, simulating a situation of risk of collision, so that, if the ship is allowed to transit towards its destiny, gradually lowering the revolutions, there is an interesting opportunity to develop a method of application of the KERS in the maritime transport.

In such situations, the vessel, thought of as a power plant, can take advantage of these speed reductions in different navigation circumstances and that sometimes lead to significant times due to inertia.

The sum of all maneuvers and times along an established route of 2 months sailing can mean days of which many hours will be spent to reduce speed or take advantage of the wind at high speeds. The tests performed must be contrasted in prototypes and navigation, where the study of the resistance to the advance has its importance since it is directly related to this system in the reduction of the speed and / or in the presence of wind, since the openings made for the conduction of the flow will not offer the same resistance as the other lines of the live work, knowing the configuration of the exits or entrances of the conduit can facilitate the formation of waves.

3.1 (a) Practical applications of the results.

The Design Index for Energy Efficiency (EEDI) in Ships is based on a series of variables among which are.

- Propelling Power
- Load capacity
- Specific fuel consumption
- Vessel Speed

High EEDI values indicate inefficient vessels and low values of this index efficient vessels. Similarly, the Operational Indicator for Energy Efficiency EEOI quantifies the environmental efficiency of a ship. EMSA (European Maritime Safety Agency) urges shipowners to adopt more efficient systems. In 2013 there is a historical fact, the daily operating costs of the vessels were higher than the daily investment costs for its construction. Fuel consumption is now the most important operating cost and one of the biggest environmental problems.

To make the economic projections of return on investment and the benefits of the installation of the proposed system, the following is taken into account, the International Energy Association (IEA) uses an energy unit, called an equivalent ton of oil (tep), whose value is equal to 107 kilocalories.

To do this, he admits that 1 *kg* of crude oil is equal to 10,000 *kcal*. Change Units:

$1\ MWh = 106W \cdot h = 108 \cdot 36\ W \cdot s = 36 \cdot 108J = 36 \cdot 108 / 4.18\ [cal] = 36 \cdot 108 / (103 \cdot 4.18)\ [kcal] = 861244\ kcal.$

As 10,000 *kcal* is equal to 1 *kg* of oil, using a rule of three, 1 *MWh* = 86.12 *kg* of oil = 0.086 tep.

Therefore, for the economic analysis, the reference of 1 *MWh* = 86, 12 *kg* of oil is of interest.
According to numerous projections since 1971, world oil reserves indicate a progressive decline in the next 75-100 years. It is accepted by a large part of the scientific community that there will be a peak independently when it takes place since it is a non-renewable source and someday it cannot be extracted. Therefore, the value of oil for the economic projection

of this research is an approximate average of prices in recent decades, taking as reference for the price of ton and type of fuel obtained from oil.

Type of fuel by Price	
IFO 380	626dollars
MDO	892dollars

For the economic projections, average values of visits to a specific port are indicated during an annual exercise, with 1 - 2 hours of approach manoeuvring [5] at the port and stopping for said vessels at a speed; 24 knots before lowering the number of revolutions in the main machine in a cruise of 330 meters in length.

To better clarify the feasibility of the project, the useful life of the vessel is considered.
Useful life: 35 years.

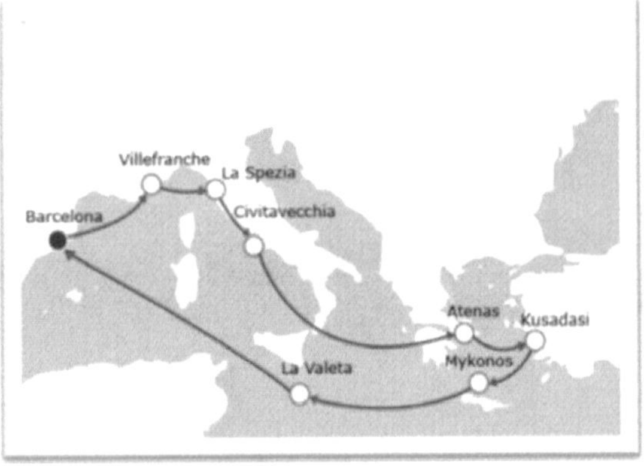

Figure 2. Typical ship route.

Premises

-Annual visits to Barcelona = 20 visits with 8 stop on route
-1MW = 86.12 kg of oil
-Price half a ton IFO 380 = 626 dollars
-Price half a ton MDO = 892 dollars
-W_A = 1131kW

-Speed 24 knots at 4 knots: 12 m/s, 8 m/s, 6 m/s, 4 m/s
-Length of 330m
-Turbine Diameters = 4m - 2m
-Seawater density; 1025kg/m^3
-Coefficient K = 0.3 (pipe without conduit, less optime)
-Useful life = 35 years

3.1 (b) Economic calculations

In these calculations, a propeller without a pipe is established as a starting premise on each side of the ship to start the manoeuvre and recover energy. A projection is made only of the precautionary manoeuvre and approach to port for large displacements, as optimal times are available for the generation of renewable electricity in its loss of speed while the revolutions of the main machine gradually decrease.

Electricity production = 20 approximations Barcelona / year x 11,272kW = 225,440kW = 225MW
Conversion: 225MW x 86.12kg oil = 19,415kg of oil = 19.42 t of oil.

Depending on the machine and the fuel used:
- Price t fuel IFO380 = 626 dollars savings = 19.42 t x 626 dollars = 12,156 dollars a year to Barcelona.

- Price t fuel MDO = 892 dollars savings = 19.42 t x 892 dollars = 17,322 dollars a year to Barcelona.

Therefore, savings are obtained depending on the propellant used:
- 12,156 dollars per year of main machine fuel.
- 17,322 dollars per year of auxiliary engines.

3.1 (c) Economic indicators

The number of pumps/turbines will depend on several key factors that need to be developed in the future. In this case, the ship will have one per side and it is assumed that there is no resistance in its advance before starting the CFD tests in the future.

Fuel IFO380 = 12,156 dollars x 2 sides = 24,313 dollars of annual savings to Barcelona.
MDO fuel = 17,322 dollars x 2 sides = 34,644 dollars of annual savings to Barcelona.

Savings in the useful life of the ship = \$ 850,955 lifetime MDO fuel = \$ 34,644 per year x 35 years = \$ 1,212,540 lifetime

A Company with 4 ships doing similar routes. Total fleet savings:

IFO380 = \$ 850,955 x 4 ships = \$ 3,403,820
MDO = \$ 1,212,540 x 4 ships = \$ 4,850,160

3.1 (d) Decarbonization transition

To know the CO2 emissions that the ship stops emitting in this manoeuvre, the following formula is used with general character in this study and included in the European directives on maritime transport [6].

$$CO2 = Ccomb \; x \; 3,17 \qquad (11)$$

Where;
3.17 is the CO2 emission factor by specific mass of fuel.
C_{comb}: Fuel consumption in Tons.

CO2 reduction

Annual CO2 cruise = *Ccomb x 3,17* = 38.84 *t* x 3.17 = 123 *t* of CO2 not emitted.
CO2 Useful Life = 123 *t* x 35 years = 4,309 *t* of CO2 not emitted.

Price *t* of CO2 (2016) = 4309 *t* x 17.61 euros = 75881 euros

Reduction fleet cruise annual CO2 = 123 *t* x 4 ships = 492 *t* of CO2 not emitted.
Useful Life = 492 *t* x 35 years = 17,220 *t* of CO2 not emitted.

Price *t* of CO2 (2016) = 17,220 *t* x 17.61 euros = 303244 euros

Example:

Extrapolating these data to the automotive sector, a passenger with his own car is estimated to have a carbon footprint of around 0.2 *kg* of CO2 per year, according to this, this passenger could travel in his own car in 1 year the distance of 647368 km to cover the 123 *t* of CO2 recovered with the progress in the precautionary maneuver and approach after 1 year the ship making its route.

Similarly, a shipping company with 4 ships has 17,220 tons not emitted in 35 years after being recovered from the sea, which could be reused in 500 plane trips, with an average footprint of 0.5 kg of CO2 per passenger, covering 76,540 km in one year.

4. KERSSHIP: PUMPED/TURBINED REVERSIBLE HYDROELECTRIC PLANT.

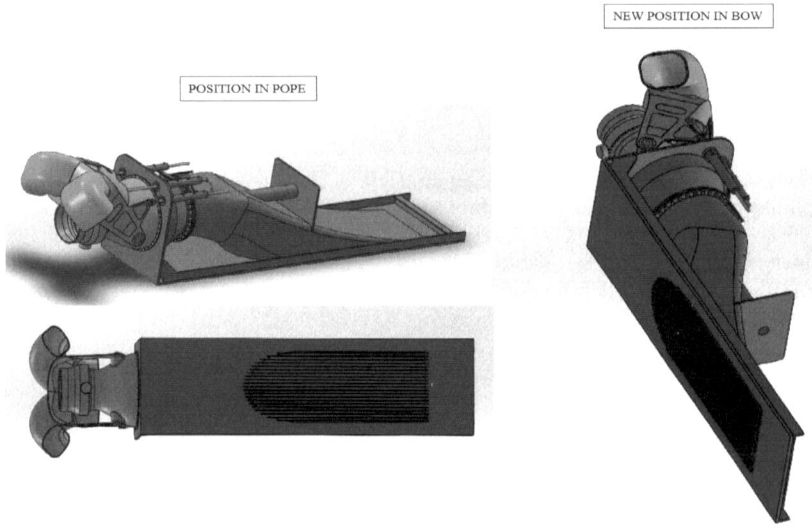

Figure 3. Waterjet Rolls-Royce.

Figure 3 shows a typical water jet. If it is turned on one side, it can adapt to the hull and function as a generator if a good design and sizing adjustment is made. It remains to solve the impeller outlet from the side, or from the keel, which will be the one that provides the thrust for the side stand. The impeller outlet carries a rotating part restricted to one degree of freedom to move around the vertical axis as many degrees as possible to expel the jet 180º from aft to bow direction.

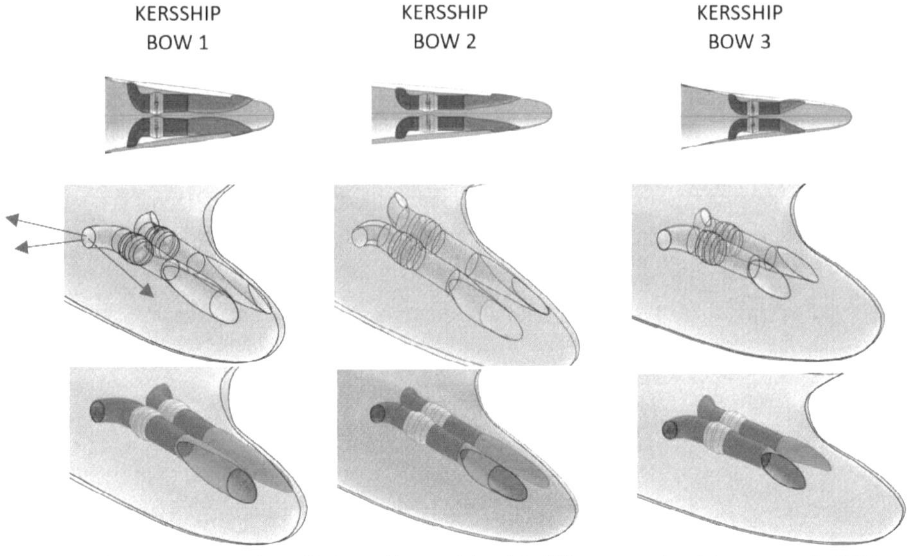

Figure 4. Preparation of 3 Kersship models with different suction inlets to test in CFD.

The Kersship system replicates the automotive KERS system, taking advantage of favorable navigation conditions to use regenerative braking. The difference is that instead of being the wheels of the car that transmit the torque to the motor shaft, it is the bow thrusters that generate electricity, and in addition, they will continue to perform the same propulsion function. For this design, it is necessary to select the machine that performs said function, within the desired requirements and for this specific configuration, the rotary electric hydraulic machines are adapted.
Specifically, the turbine/pump of pumped-reversible power plants is of interest. In the side outlet, figure 4 indicated with arrows, there is a rotating outlet that propels the water jet at various angles: forward, backward, or perpendicular to the length. Without going into the machine in depth, a generator is a rotary electric and its main function is to transform mechanical energy into electrical energy. Electric generators are divided into two large groups: alternators (alternating current) and dynamos (direct current).

5. TAILWIND AND SAILS.

To find high efficiency during navigation when it comes to recovering kinetic energy, both the positive incidence of the wind on the stern surface or under the use of sails when weather and navigation conditions require it is relevant, finding high efficiencies in reducing consumption by having extra thrust of 5 knots in traditional boats and 11 knots for boats with a better aerodynamic profile, according to references [7][8][9][10], the sterns of a cruise ship have enough width to take advantage of this thrust in the tailwind condition.

In fact, with the current systems installed on board for the detection of climatic conditions, the captain can anticipate in advance the most optimal route to take advantage of that extra impulse, placing the ship in the best position to receive the wind from the stern or at favourable angles. Favourable wind angles, or force vectors, that do not impinge on the stern surface can be captured by rigid, rotating and folding sails seeking to chase currents and increase thus the speed of the ship being able to lower revolutions in the main machine. To find the thrust T ($Kgfr$) on the stern of the ship and on the sail surfaces, the following formula [8][9] is used, for a given wind angle.

$$T = 0.5 \cdot \rho a \cdot V^2 a \cdot A \cdot Cx$$

(11)
Where,
ρa, is the air density. ($kg \cdot s/m^4$)
V^2a, is cube velocity. Apparent Wind Speed: Va (m/s)
A, is total sail area. Area of wing sail (V^2)
Cx, is the thrust coefficient (variable parameter according to the apparent wind angle β).

Therefore, the thrust produced by the wind provides gains in speed, of up to 11 knots in the best condition, this makes it possible to "requisition" about 3 or 5 knots and sustain losses due to resistance to movement due to openings or roughness which can appear when the hull has an opening in the bow and on the sides. On the other hand, most of the remaining knots (5 or 7 knots) are derived to lower the revolutions of the main machine and reduce consumption. While the wind maintains almost the same speed of the boat at 20 knots at lower revolutions. At this point, that loss is of interest after requiring 3-5 knots to maintain the drag mentioned above, because this restriction prevents the boat from increasing speed without extra expense and covering the route in less time thanks to the wind or use those 11 knots directly to lower revolutions to continue sailing at 20 knots and reduce consumption even more [12] while maintaining the schedule established in the charter. This is explained, for example, if a ship sails at 20 knots in a normal way with average speed and has an extra push for two hours of 7 knots provided by the wind accelerating the ship, then, if 4 knots of the 7 total knots are used to reduce the revolutions of the main engine, maintaining the initial 20 knots and the other 3 remaining knots will be used to use the generator/bow thruster to recover kinetic energy at 20 knots for two hours, this involves testing the best option, well, if the main machine is a diesel-electric engine, that energy recovered from the sea could be directly injected and the revolutions continue to drop while the bow thrusters work as generators on demand from the main machine, maintaining cruising speed.

The power generated at 20 - 24 knots, in this hypothesis, can be injected directly into the main engine downwind without losses by resistance and, this situation, could be more favorable in terms of efficiency than deriving the entire 7 knots to sail at 27 knots or allocating them only to lower revolutions. Other alternatives [13] to take advantage of this energy is to produce hydrogen, or used to cover peaks of high demand of the main machine, for example, with gusts of wind against the advance or when starting the march leaving the port, etc.

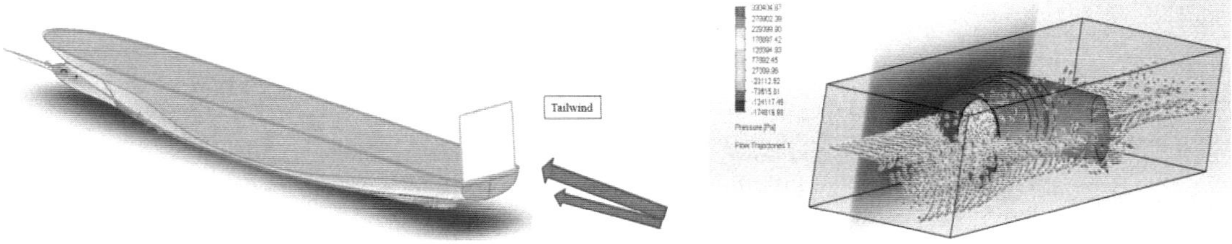

Figure 5. Representation of the Tailwind and CFD basic test in duct. Own elaboration

7. CONCLUSIONS

- From the analysis of the available information, a series of empirical project criteria have been compiled or deduced quantitatively (formulas, basic calculation, and graphs), either, in the form of qualitative recommendations. Explanatory examples, descriptions and designs of the applied technology are provided.

- It is also necessary the continuation of theoretical studies on the phenomena that produce the openings in the hull to produce clean energy, viscosities, suction, propulsion effect during navigation in manoeuvres with loss of speed, while remaining anchored or being anchored. in tow, in conditions with stern wind and high speed, etc.

- The indicators and economic projections show that the results obtained in this document in decarbonization could be improved between 15 and 20% under optimal designs of the reversible pump/turbine together with good planning of the thrusts produced by the tailwind and captured by the sails systems.

- The development of this complete project requires a lot of computational resources and human capital in different disciplines to carry out an optimal design of the proposed system.

- The first CFD tests were carried out in different operations and speeds, without a propeller, showing important design errors in the aspiration when carrying out the tests without the hull surface, which would allow a better transition of the fluid. Estos ensayos iniciales serán importantes cuando se desarrollen test más complejos y entre en juego el NPSH.

- Climate change, fuel shortages, favorable legislation should favor disruptive innovation in order to optimize processes and systems.

8. ACKNOWLEDGEMENTS

To my godmother, Nina.

9. REFERENCES

1. Hyundai Heavy Industries Co. Ltd. "Ship having micro hydro turbine". Patent Number: KR101400595B1. 2013

2. Urzhumov, Yaroslav. Effortless Sailing with Fluid Flow Cloak. Duke University. 2011.

3. Fernández Diez, Pedro, "Corrientes Marinas II", Page 34-35, Retired profesor. 2008.

4. Grûnewald, Antonio Ramón. "Estudio de la generación energética con turbinas hidráulicas en las corrientes marinas". Renewable Energy Master. Pages 84 to 89. 2012.

5. International Maritime Organization, ANNEX 6 RESOLUTION MSC.137(76) STANDARDS FOR SHIP MANOEUVRABILITY, Resolutions IMO, 2002.

6. EMEP/EEA air pollutant emission inventory guidebook 2019. Technical report No 13/2019; 1.A.3.d Navigation (shipping) 2019. European Environment Agency. 2020.

7. N. Bialystocki, D. Konovessis, "On the estimation of ship's fuel consumption and speed curve: A statistical approach", Journal of Ocean Engineering and Science, p.162, 2016.

8. K. Ouchi, K. Uzawa, and A. Kanai, "Huge Hard Wing Sails for the Propulsor of Next Generation Sailing Vessel," no. June, pp. 1–5, 2011.

9. K. Ouchi, K. Uzawa, A. Kanai, and M. Katori, "Wind Challenger the Next Generation Hybrid Sailing Vessel," no. May, pp. 562–567, 2013.

10. Resolution MEPC.213(63), "Guidelines for the development of a ship energy efficiency management plan (SEEMP)", Annex 9- page 11, 2012.

11. L. Cachaza, G. Pérez, R. Villa, J. Fraguela, "Reducción del consumo de combustible mediante velas rígidas y paneles solares". COPINAVAL. 2015.

12. Marcus Bentin, David Zastrau, Michael Schlaak. "A new routing optimization tool-influence of wind and waves on fuel consumption of ships with and without wind assisted ship propulsion systems". Procedia. 2016.

13. International Maritime Organization Directrices de 2013 para el cálculo de los niveles de referencia que se utilizarán para el índice de eficiencia energética de proyecto (EEDI) para buques de pasaje dedicados a cruceros con sistemas de propulsión no tradicionales (resolución MEPC.233(65))

9. AUTHORS BIOGRAPHY

Gerardo García Lage holds the current position of PhD student at University of A Coruña. He is responsible for carrying out the research until the moment of defending the doctoral thesis. His previous experience includes application and granting of Invention Patent with world novelty in all claims, patent grant in 2019, awarded by the University of A Coruña in port sustainability by TFM 2019, awarded by the AINE and the COIN in maritime efficiency at the Naval Engineering Congress Barcelona 2016, The doctoral research was selected in competitive competition by the European program Interreg Galicia-Norte de Portugal as a project of interest for the decarbonization of maritime transport, a program that ended in March 2022, articles published in magazines of the sector at a national level in Spain, speaker at 4 conferences on efficiency and R&D in the naval and port sector, in regional and national media published news of some of my awards and achievements.

INVESTIGATION OF WATERJET EFFLUX NOISE WITH NUMERICAL TECHNIQUES

S Dhavalikar, J Ramkumar, P Kumar CH and **A Chaturvedi,** Indian Register of Shipping, India

SUMMARY

The URN resulting from a waterjet propulsion (WJP) is to be accurately estimated for stealth considerations. Such WJP systems are designed to operate either with a submerged jet (wet transom) or with a free surface jet (dry transom). The URN due to the operation of WJP system can be broadly categorized as pump noise and nozzle efflux noise. Pump noise includes mainly the cavitation noise and the tonal mechanical/hydrodynamic noise, whereas the nozzle efflux noise is due to jet impact and bubble entrainment. The present study attempts to implement advanced numerical techniques to estimate URN due to the nozzle efflux of a WJP in both wet and dry transom modes. Large Eddy simulations (LES) are carried out to simulate the hydrodynamics of turbulent multiphase flow behind both submerged and free surface water jets. The impact of water jet on free surface and subsequent bubble generation rate is captured using multiphase volume of fluids (VOF) technique. The impact noise is estimated by directly recording the fluctuating acoustic pressure at certain field points during the simulation. A blob detection algorithm is implemented to compute the bubble generation rate. A mathematical model to compute the bubble pulsation noise from bubble generation rate is validated and implemented. The estimated URN levels in wet and dry mode are presented and compared. Key inferences are presented which could be useful in further investigation of URN from water jet of stealth vessels, as well as to investigate bubble noise from propeller cavitation, air lubrication etc.

NOMENCLATURE

μ	Dynamic viscosity of water (N s m^{-2})
ρ	Density of water (kg m^{-3})
γ	Polytropic index of gas (-)
ω_0	Oscillatory resonance frequency of the air bubble entrapped in the water (rad s^{-1})
c	Velocity of sound in water (m s^{-1})
P	Pressure (N m^{-2})
P_0	Total sum of hydrostatic and atmospheric pressure (N m^{-2})
R_0	Initial radius of the bubble (m)
CFD	Computational Fluid Dynamics
LES	Large Eddy Simulation
PSD	Power Spectral Density
SPL	Sound pressure level
URN	Underwater Radiated Noise
WJ	Waterjet
WJP	Waterjet propulsion

1. INTRODUCTION

A typical WJP system is composed of an inlet channel, a rotating pump impeller, a guide vane, an outlet nozzle and a steering/reversing mechanism. During operation, the rotating pump impeller accelerates the water flow directed from the inlet channel towards the outlet nozzle through the guiding vane. This accelerated water effluxes out of the nozzle as a jet, which is directed by the steering / reversing mechanism. This jet results in a turbulent flow in the wake region behind the nozzle. When the nozzle is submerged the jet directly mixes with the flow in the wake region of the vessel. If the submerged jet is near to the free surface, then this jet impingement results in a disturbance eventually leading to air bubble entrainment. When the nozzle is at the level intersecting or above free surface, the jet impacts over the free surface leading to a greater turbulence in the wake region. Therefore, the air bubble entrainment is greater in case of a free surface jet.

Mechanical/hydrodynamic phenomenon during the operation of the water jet propulsion results in URN, which is of significant interest for stealth vessels. This URN is to be accurately estimated at the design stage to improve the stealth characteristics of the vessel and its propulsion system. An accurate estimation of URN at various operational conditions would play a significant role in success of a stealth mission.

The primary sources of URN from a WJP can be categorized as the pump noise and the water jet efflux noise. The pump noise is mainly due to the impeller interaction with the flow which result in cavitation noise and/or tonal noise. The other category of noise source is jet efflux.

When WJP system is operated in wet transom mode, i.e., the outlet nozzle completely submerged under water, the efflux mixes directly with the flow astern the vessel. In this mode of operation, the efflux URN is low as there is no free surface impact and also due to significantly lower air bubble entrainment. When the submerged jet efflux is near the free surface, then there could be an increase in the air bubble entrainment rate, resulting in higher URN.

When in dry transom mode of operation, the water jet from the nozzle effluxes above (or at) the free surface level and impacts the free surface, causing noise. The impact also causes turbulence in flow astern of the vessel, leading to air entrainment and subsequent bubble generation. The generated bubbles initially undergo volumetric oscillation leading to a bubble pulsation noise. The mechanism of volumetric oscillation can be attributed to surface tension and high compressibility of air relative to water. Surface tension always acts minimizing the bubble surface area trying to make it spherical in shape. During their formation the bubbles are imparted with an initial excitation which is manifested as a difference between bubble internal and external pressures. When the internal pressure is more than the external pressure, then the bubble volumetrically expands resisting the surface tension. As the bubble expands the internal pressure reduces below the external pressure. When the bubble external pressure is more than internal pressure the bubble contracts in volume, while surface tension keeps it spherical. Therefore, the volume of the bubble oscillates around a mean volume which is termed as bubble pulsation. This bubble pulsation exponentially decays due to thermal, viscous and radiation damping.

The radius of the bubble fluctuates around a mean radius (R_0) as a decaying sinusoidal wave. Half the difference between maximum radius and minimum radius in a cycle of oscillation is the amplitude of bubble radius. The ratio of initial amplitude of pulsating bubble radius ($R_{0\epsilon}$) to mean radius(R_0) can be used as a quantification of the excitation. This ratio $\frac{R_{0\epsilon}}{R_0}$ is higher for a bubble subjected to greater excitation during its formation. The bubble volumetric pulsation causes high frequency pressure fluctuation in the field which is observed as bubble pulsation noise. This noise exponentially decays as the pulsation decays.

The objective of the present study is to numerically predict URN from a water jet propulsion operating in both wet transom and dry transom mode of operations. The scope is limited to study of noise resulted only due to the water jet efflux, but not the pump noise. Therefore, only the physics of free surface flow and acoustics abaft the water jet nozzle are modelled. The net URN is considered as a superimposition of impact noise and bubble pulsation noise.

In order to establish a methodology to achieve the said objective relevant literature is surveyed. The literature on the studies pertaining to the water jet noise is limited. So generic literature pertaining to water droplet impact on free surface, underwater air bubble formation, pulsation and noise estimation is studied and presented in section 2. Following the literature study, a methodology to estimate URN is formulated and implemented. The methodology implemented for CFD simulation of water jet efflux is presented in section 3. Section 4 presents the formulation used to compute fluctuating pressure to estimate impact noise. Section 5 presents the formulation to estimate bubble pulsation noise. The results of free surface water jet simulation and submerged water jet simulation are presented in section 6 and 7 respectively. Section 8 presents inferences from the present study and concludes.

2. LITERATURE REVIEW

The literature on water drop impact on the water surface and relevant acoustic emissions is relevant in context with WJ noise. Noise due to water drop impact on the water surface and the bubble(s) generated due to this impact are discussed in literature. The relevant physics and the mathematical models developed are applied to the problems for acoustic emissions from e.g. stream flow, rain drop, gas leakages from pipelines underneath the ocean. The WJ noise problem is similar in nature thus the models available in literature can be applied to WJ noise estimation.

The first thorough investigation of the sounds produced by drop impacts was made by Franz [1]. The main sources of underwater sound from a splash that he discovered were i) the impact and passage of water droplet through free surface leading to establishment of flow, ii) resonance vibrations of the body if it has rigidity and iii) volume pulsations of bubbles of air in the water. Franz observed that bubbles were not produced by every drop and that their occurrence was random. Further experimental investigations were made by Pumphery et al [2] and their findings provided evidence for the theory that the 14 to 16 kHz spectral peak is caused by the ringing of bubbles entrained in the water by the drop impact process. The study also revealed that the water drop impact consists of two distinct parts i) sharp leading edge which is true radiated sound and ii) pressure variation following the leading edge. The second part is purely hydrodynamic effect noticed in the vicinity of the impact and it is not true sound at all. The true initial impact pressure P_i is proportional to the drop diameter to the power 2.1 ± 0.6. P_i is also proportional to the impact velocity V_i to a power between 2.5 and 3 which was in line with Franz [1]. Continuing the similar investigations with modern tools, Prokhorov & Chashechkin [3] investigated water drop falling modes where they noted arbitrary cavity shape gradually achieve elliptical and spherical shapes in equilibrium. Phillips et al [4] investigated the reverberation effects in the tank while studying the water drop sounds. The air borne sound fields in addition to underwater sounds were also investigated.

Doel [5] developed a physically based liquid sound synthesis methodology. The fundamental mechanism for the production of liquid sounds was identified as the acoustic emission of bubbles. Bubbles are formed when the water surface causes air to be trapped in the water, or when air is injected in the liquid by a nozzle, or by blowing with a straw in a glass of water or some such mechanism. Bubble formation is usually accompanied by an energy injection into the bubble at creation time. This can happen for example when a surface wave breaks and traps bubbles, when a cavity created by a falling droplet collapse, or when a parent bubble breaks up and surface tension excites the child bubbles. After formation, the bubble emits a sinusoidal sound which decays as energy is dissipated. If the bubble survives long enough, this is all that happens. The bubble may also break up into smaller bubbles, which are excited by surface tension forces at the time the bubble breaks up. This process would result in a pair of bubble sounds immediately after the original bubble sound. The acoustic mechanism responsible for bubble sounds is volume pulsation, which was first correctly identified by Minnaert [6]. One way to visualize this is to think of the bubble as a small compressible region added to the incompressible fluid, allowing it to oscillate. Another viewpoint is to consider the bubble as a source of underwater sound which propagates through the liquid. The physics of a single spherical bubble is well understood, however, in reality, bubble sounds are almost never heard in isolation. Even a relatively simple event such as a single drop falling in water generates a complicated sequence of bubbles and secondary droplets. The solution to the problem would be to synthesize the sound made by water by numerically solving the compressible Navier-Stokes equation for water and air at audio rates, i.e., with time steps of about 1/44100 s [5].

Leighton and White [7] provided a model to calculate the acoustic pressure radiated by a single bubble pulsating at natural frequency ω_0 at time t and a distance r (far field). Li et al [8] studied bubble formation dynamics using acoustic techniques discussed in [7]. The study was primarily towards the acoustics due to under-sea gas leakages. It was found that the gas discharge rate has strong impact on the sound of bubble formation. The choice of nozzle imparts some influence on the bubble formation. Li et al [9] estimated bubble generation rate as a reverse problem of the model proposed by Leighton and White [7].

Recently Luo et al [10] presented a co-simulation method combining the CFD techniques and computational aero acoustics to predict the URN of submerged high pressure jet propulsion.

In the present paper authors have used the model proposed by Leighton and White for bubble noise. The bubble generation rate is computed based on CFD based simulations. The experimental results of Li et al [9] for underwater gas bubbles are used for initial validation of bubble noise. For WJ simulations impact and bubble noise are distinguished. Impact noise is approximated based on the oscillating pressure in the computational domain and bubble noise is estimated based on mathematical model.

3 CFD FLOW SIMULATION

In order to predict the impact noise, the turbulent multiphase flow due to the water jet nozzle efflux is first simulated utilizing the CFD solver StarCCM+. Figure.1 shows the schematic of computational domain considered for the free surface WJ flow simulations. In case of submerged jet, the jet centre is kept at d times below free surface where d is the diameter of the jet. Half breadth and the depth of domain are kept sufficiently large (more than 10d) in each simulation.

Table 1. Parameters used for WJ simulation.

Parameter	Value	units
Jet diameter, d	0.5	m
Flow velocity at inlet	2.0	m/s

Table 1 shows the parameters considered for simulation. Flow velocity of 2m/s at the nozzle inlet is considered. The nozzle action accelerates the flow and the velocity achieved at nozzle outlet (for considered configuration) is ~8m/s.

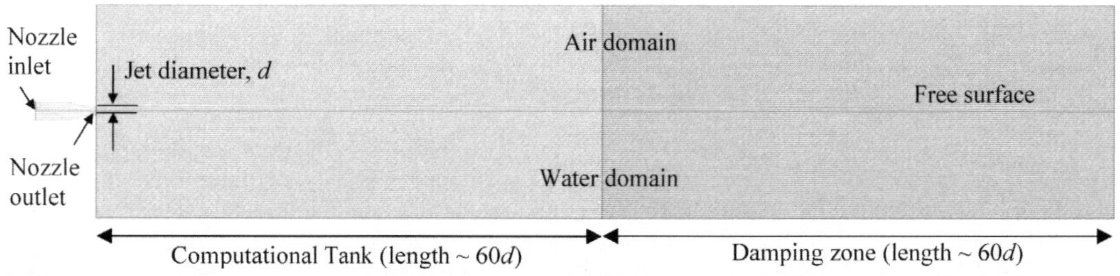

Figure 1. Schematic of computational domain for waterjet simulations

The fluctuating pressure field, velocity field and the distribution of air water phases in the domain are predicted and recorded in time. The recorded time history of pressure at a given point in the field is a superimposition of the low frequency hydrodynamic pressure and the high frequency acoustic pressure, later being the effect of URN. From the recorded air water phase distribution the number of bubbles and their corresponding radii are computed at each time step. The bubble generation rate for various bubble radii is computed. A mathematical model to estimate the bubble pulsation URN from the bubble generation rate is established, validated and implemented.

CFD techniques are used to simulate the water jet efflux from the nozzle, and the resultant flow astern of the vessel. The interest is limited to the flow in the outlet nozzle and a wake region behind the vessel where the jet impingement effects are significant for the present study. The simulation domain is composed of two regions, one constituted of a portion of the conical nozzle and a rectangular prism representing the flow region behind the hull (blue zone in figure 1), the other region is also a rectangular prism behind the first region (red zone in the figure 1). The dimensions of the regions are chosen based on two aspects: 1. capturing the effects of jet efflux and 2. Mitigating the numerical reflections from the boundaries. The second region behind the original flow region is considered mainly to avoid boundary reflections.

The computational domain comprises of two immiscible phases namely air and water. The interface is resolved using Volume of Fluids (VOF) technique, and High-resolution interface capturing (HRIC) model. This multiphase flow is defined by field variables: Pressure, velocity and volume fraction of water and air distributed in the domain. These flow variables are initialised such that an undisturbed free surface is set at given level inside the domain. The boundary conditions are set such that a stream of water jet effluxes out at constant velocity from the surface representing the nozzle outlet. All other surfaces are set to either pressure outlets or velocity inlets with the numerical values simulating a no flow or free flow conditions. Table.2 shows the boundary conditions and initial conditions. The flow is modelled to be turbulent, with air and water being compressible. Hence the flow is also affected by temperature, and this has to be solved. The compressible continuity equation, Navier-stokes equations and the energy conservation equation are solved in time to simulate the flow. Large Eddy simulation scheme is used to resolve the larger eddies, whereas the smaller eddies are modelled using Smagorinsky Subgrid Scale model. Implicit filtering of smaller eddies is used numerically. The computational grid determines the scales of eddies that are filtered out. Implicit filtering takes full advantage of the grid resolution and is, in general, computationally less expensive. In contrast the spatial filtering removes the smaller eddies—associated with higher frequencies—and thereby reduces the range of scales that must be resolved, in the present study the eddy sizes below 0.0005mm are filtered out.

Table 2. Domain Boundaries and Boundary Conditions

Boundary Name	Boundary Type
Nozzle Inlet	Velocity Inlet
Nozzle Boundaries	Wall
Tank Walls	Wall

Tank Sides	Symmetry
Tank Bottom	Wall
Tank Top	Opening (Relative Zero Pressure)
Tank Outlet	Pressure Outlet (Hydro-Static)

A blob detection model is used to detect the bubbles and their radii at each time step. The distribution of bubble generation rate vs bubble radius is obtained from the blob data from consecutive time steps.

4 WJ EFFLUX NOISE

Figure 2 shows the flow captured during WJ impact following the CFD method as briefed in previous section.

Figure 2. WJ flow a) bubbly flow (activating blob model), b) flow velocity on the free surface

It is noted that bubbly flow is generated, and the bubbles are formed only after the impact of water on the free surface. The noise due to impact and the bubbles is to be treated separately.

4.1 IMPACT NOISE

The mass of water in the form of jet impacts the free surface. The impact pressure of the order of $1/2\rho V^2$ can be expected in the impact region, where V is the velocity of the flow. This pressure is purely hydrodynamic effect. This impact causes the minor fluctuations in the pressure in overall continuum and these fluctuations radiate far away from the source. The radiated fluctuations are relevant for acoustic investigations. The fluctuating pressure is nothing but the difference between the mean pressure in the fluid flow (fluid domain) and the instantaneous pressure at a given location in the domain. Figure 3 shows the fluctuating pressure measured in the way of the jet (nearly 6m away from jet outlet). It can be noted that the peak pressure magnitude is of the order of 8000 Pa which is in the proportion of the flow velocity (Approx. 4m/s, Ref figure 2 b). Thus, these fluctuations can be considered purely due to the hydrodynamic effect. For acoustic investigations the pressure fluctuations away from the jet turbulence should be captured, e.g., as shown in the figure 4.

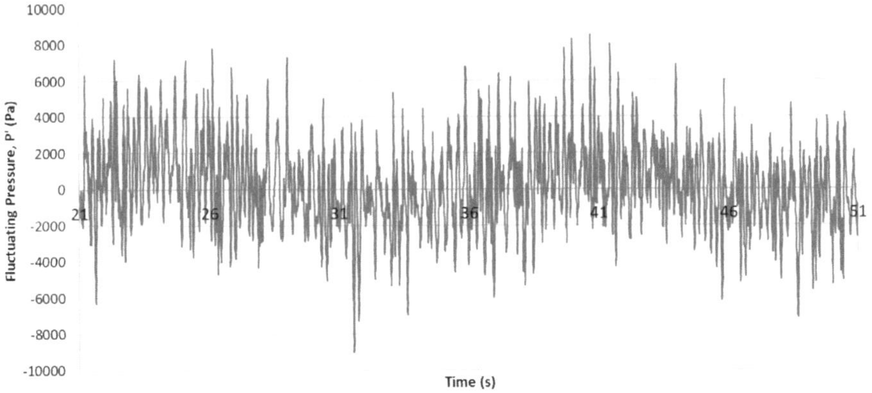

Figure 3. fluctuating pressure in the way of the jet

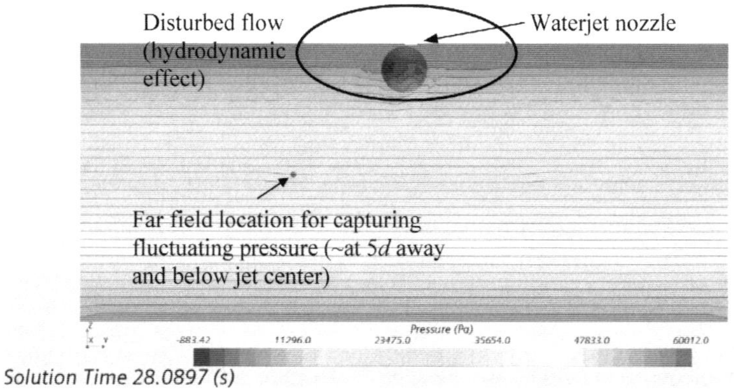

Figure 4. Fluctuating pressure measurement location for acoustic investigation

Figure 5 shows the fluctuating pressure variation at the far field. It can be noted that the peak fluctuation is 519 Pa which in terms of jet velocity (V = 8m/s) is $V^{3.0}$. This finding is in-line with the Franz [1] which states the impact pressure relevant for acoustics is proportional to the impact velocity $V_i$to a power between 2.5 and 3.0. Further, presently authors consider finding the acoustic power relevant for WJ impact is not feasible. Hence, the fluctuating pressure signal is directly converted to sound pressure level with suitable reference pressure (1µPa) to find the contribution due to impact.

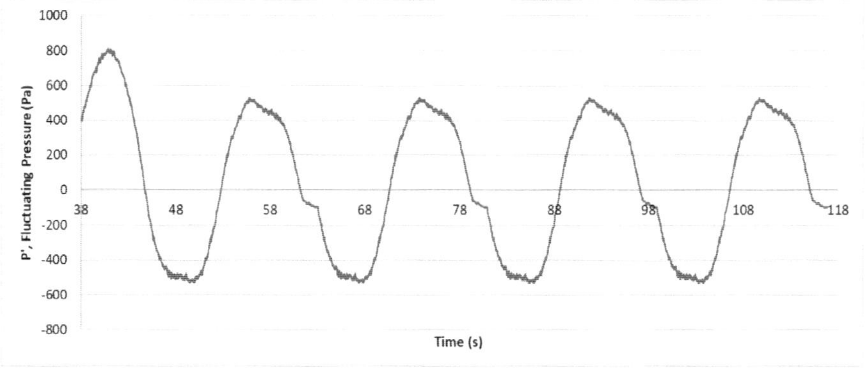

Figure 5. Fluctuating pressure at far field for Free surface jet

4.2 BUBBLE NOISE

With the impact of water, the cavities are created in the fluid domain which is similar to rainwater drop impacts. These cavities are filled with air from free surface. The bubbles thus generated start vibrating due to compressibility and surface tension effects. The amplitude of vibrations is very small however the number of bubbles generated can be very high depending upon the jet velocity and overall flow physics. The vibration of bubbles is another source of noise in the fluid domain.

The sound pressure spectrum of the acoustic emission of bubble generation is given as below [7] -

$$|P(\omega, R_0)|^2 = \left(\omega_0^2 R_0^3 \frac{\rho_0}{r} \frac{R_{0\varepsilon}}{R_0}\right)^2 \times \frac{4\left((\omega_0\delta_{tot})^2+4\omega^2\right)}{\left((\omega_0\delta_{tot})^2+4(\omega_0-\omega)^2\right)\left((\omega_0\delta_{tot})^2+4(\omega_0+\omega)^2\right)} \quad (1)$$

Where, ω_0 can be found using Eq. 2 (Minneart formula [6])

$$\omega_0 = \frac{1}{R_0}\sqrt{\frac{3\gamma P_0}{\rho}} \quad (2)$$

γ = polytropic index of gas (air in present case) = 1.4

The decay of the pressure induced by the oscillating bubble is determined by the dimensionless damping constant δ_{tot} defined as

$$\delta_{tot} = \delta_{th} + \delta_{vis} + \delta_{rad} \quad (3)$$

The thermal damping constant δ_{th}, the viscous damping constant δ_{vis} and the radiation damping constant δ_{rad} are given as below

$$\delta_{th} = 4.41 \times 10^{-4}\sqrt{\omega_0/2\pi};\ \delta_{vis} = \frac{4\mu}{\rho\omega_0 R_0^2};\ \delta_{rad} = \frac{\omega_0 R_0}{c}$$

The power spectral density S(ω) of far-field sound can be calculated by using [3]

$$S(\omega) = \int_0^{\infty} D(R_0)|P(\omega, R_0|^2 dR_0 \qquad (4)$$

Here, the product $\{D(R_0)dR_0\}$ indicates the number of bubbles per second produced that have radii between R_0 and $R_0 + dR_0$. Bubble generation rate $\{D(R_0)dR_0\}$ can be calculated experimentally or utilizing the numerical technique (CFD simulations). Present study discusses numerical studies for finding the bubble generation rate as required for solving Eq (4). Another important parameter required to be known is the ratio $\frac{R_{0\varepsilon}}{R_0}$ in the Eq (1). This ratio indicates the bubble vibration amplitude which gives rise to the noise. Discussions on estimating this ratio can be found in [9]. General recommendation given by Leighton and White [7] is for bubbles of radius up to 1mm, this ratio is of the order of 1e-5 and for bubble radius varying between 1 – 5mm the ratio can be taken to be 3.74e-4. [8] suggests this ratio as a function of air flow rate, typically for the cases where pressurized air is passed into the liquid. The following Eq (5) is relevant in this regard

$$\frac{R_{0\varepsilon}}{R_0} = 0.0446e^{-0.4841Q} + 0.0183e^{-0.0129Q} \qquad (5)$$

where Q is the flow rate in l/min.
It can be noted that sound pressure varies in the proportion of square of the bubble vibration amplitude $\frac{R_{0\varepsilon}}{R_0}$. The resulting PSD can be drastically different than expected depending upon the choice of $\frac{R_{0\varepsilon}}{R_0}$.

4.2(a) Bubble Noise Validation

Chen et al [9] carried out experiments to predict bubble size distribution resulting from air discharged through single orifice and measuring acoustic emissions. Air was discharged at a particular flow rate through an orifice into quiescent water contained in a large tank. Acoustic measurements were made at different locations in the tank using a single hydrophone and the experiments were carried out with two orifice sizes. For 2mm orifice the air flow rate of 0.08333 l/s and for 8mm orifice air flow rate of 0.3333 l/s were used. In the present paper same study with numerical simulations is carried out. Objective of the numerical assessment is to find the bubble generation rate, the vibration amplitude of bubbles and estimating the acoustic power based on the model described with Equations 1 – 4.
Figure 6 shows the snap of bubbles generated with 2mm orifice. The PSD measurement location of 2m away from the centre of the orifice (r in the Eq 1) and at a depth of 2m from free surface is considered for calculations. Simulations with and without surface tension are carried out. It is clear from Figure 6 that air bubbles are captured correctly when surface tension is considered.

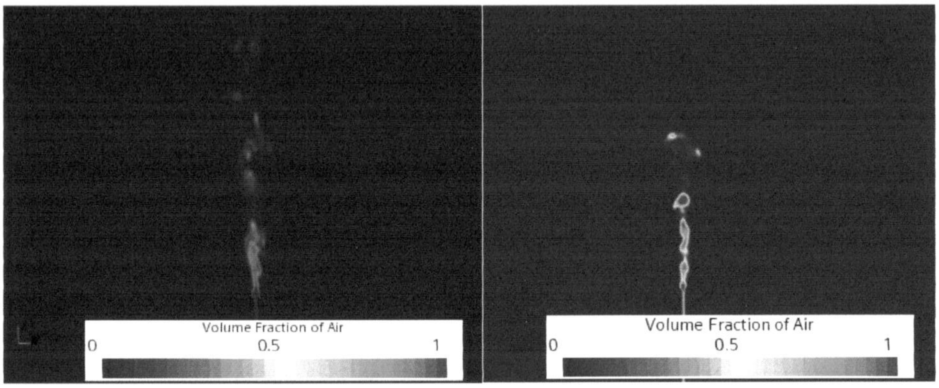

Figure 6. Numerical simulation of underwater air flow through 2mm orifice – without surface tension (left) and with surface tension (right).

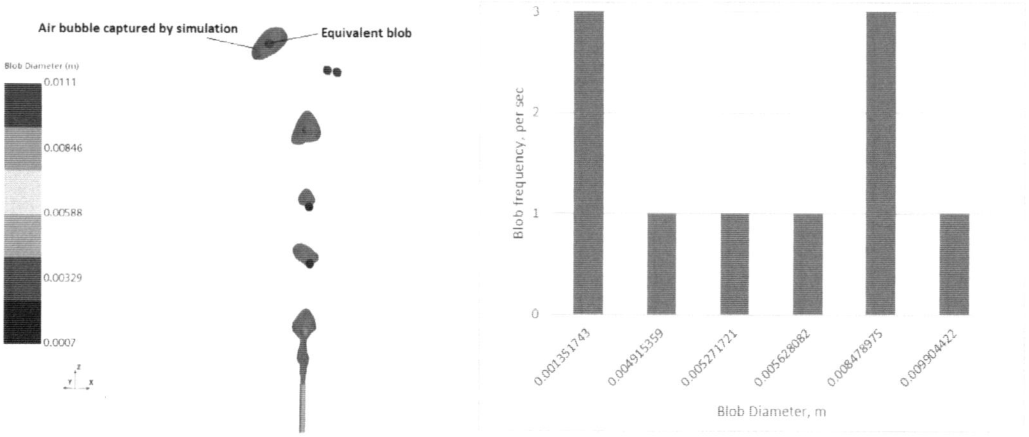

Figure 7. Converting air volume to blobs (left) and finding bubble generation rate (right).

Figure 7 shows the bubble generation rate computed from numerical simulation. Each irregular shaped air volume is converted to equivalent spherical shape and considered as bubble for finding bubble generation rate. The colour of the blob in Figure 7 indicates the size of the blob (Ref legend of the figure). For a converged solution the bubble generation rate i.e., the number of bubbles generated per second are found. Amplitude of vibration of each blob can be estimated by tracking the blob over time and space in the computational domain. Each blob is tracked by unique index. However, if the blobs coagulate to form bigger blob or disperse into small size blobs, the indices are reassigned. Accordingly, the blob indices are tracked for smaller duration to ensure unique blob is tracked. Figure 8 shows blobs of ~1mm monitored over time.

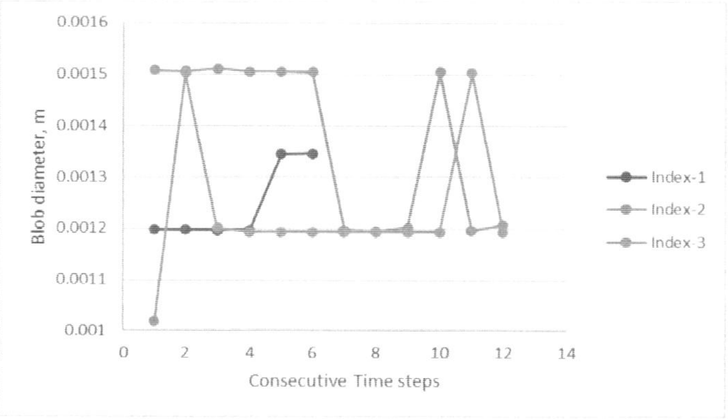

Figure 8. Estimating bubble vibration amplitude.

To find the ratio $\frac{R_{0_\varepsilon}}{R_0}$ averaging over time as below is considered

$$R_{0_\varepsilon} = \frac{1}{n}\sum(R_{0_{j+1}} - R_{0_j});\ R_0 = \frac{1}{n}\sum(R_{0_{j+1}})$$

Where n is the number of time instances (j) considered (~ 10 in Figure 6). For each blob index similar procedure is followed. The results are tabulated in Table 3.

Table 3. Estimating bubble vibration amplitude

Index	R_{0_ε}	R_0	R_{0_ε}/R_0
1	2.96e-5	0.001246	0.0237
2	-2.73e-5	0.001378	-0.0198
3	1.5848e-5	0.001231	0.0128

The negative numbers in Table 3 indicate the bubble compression whereas positive numbers are bubble expansion for the considered time instances. It is noted that the non-dimensional amplitude of vibration ($\frac{R_{0_\varepsilon}}{R_0}$) for bubbles of ~1mm diameter is of the order of 0.0188 (average of three cases). For other bubble sizes the ratio could not be verified, however, it is

expected that the vibration amplitude increases with bubble radius (R_0). Using Eq (5) the ratio R_{0_ε}/R_0 is equal to 0.021and is independent of bubble size.
The average bubble size for the generated bubbles (Figure 4) is more than 5mm. Accordingly, for calculating PSD (Figure 7) the ratio ($\frac{R_{0_\varepsilon}}{R_0}$) can be considered as more than 0.021 (say 0.025 as conservative) for all bubble sizes. Effect of varying ($\frac{R_{0_\varepsilon}}{R_0}$) is also shown in Figure 7. The background noise reported in experimental results [9] is subtracted for comparison for bubble PSD.

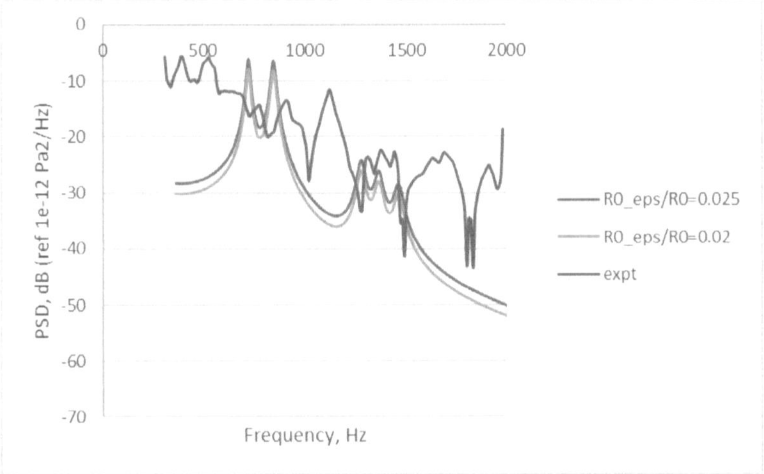

Figure 7. Comparison of calculated and measured bubble PSD at 2m.

5. RESULTS AND DISCUSSION

5.1 CONVERSION TO SPL

Following the methodology as discussed in Section 3 and 4 the computations are made for free surface and submerged jet to find the sound pressure level at far field. In general SPL at 1m from the source is reported. The correction is made by using $20\log(r)$ where r is the far field distance (measurement location).
Further for bubble noise sound pressure (p in Pascal) is found by converting PSD as

$$p = \sqrt{\frac{Q\rho c P_{ac}}{4\pi r^2}}$$

Where, Q = directivity factor, assumed to be 1 (spherical), c is the speed of sound in water = 1500 m/s, P_{ac} is the acoustic power in watts. Sound pressure is converted to dB using reference pressure of 1 μPa.

5.2 EFFECT OF SUBMERGENCE

Figure 8 shows the free surface velocity plot for the case of submerged jet. Comparing it with figure 2b it can be noted that the disturbance in the flow is much reduced in comparison with the free surface jet case.

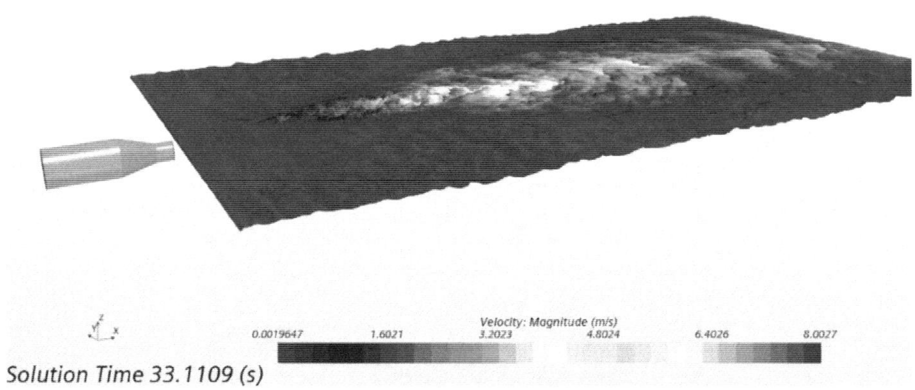

Figure 8. velocity plot for submerged jet

The fluctuating pressure (figure 9) is also reduced compared to free surface jet case (figure 5). Accordingly, the noise also gets reduced and can be seen in figure 13.

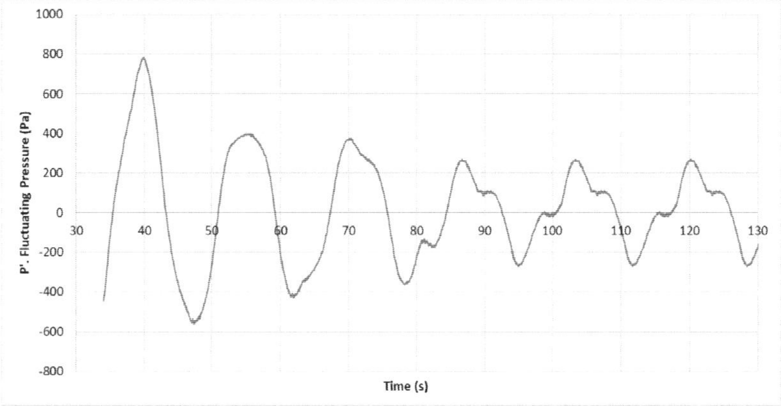

Figure 9. Fluctuating pressure signal for submerged jet case.

It can be noted that the peak fluctuating pressure in case of submerged jet falls to ~ 260Pa which is approximately $V^{2.68}$ (lesser than free surface jet case).

5.3 CORRECTION FOR BUBBLE GENERATION RATE

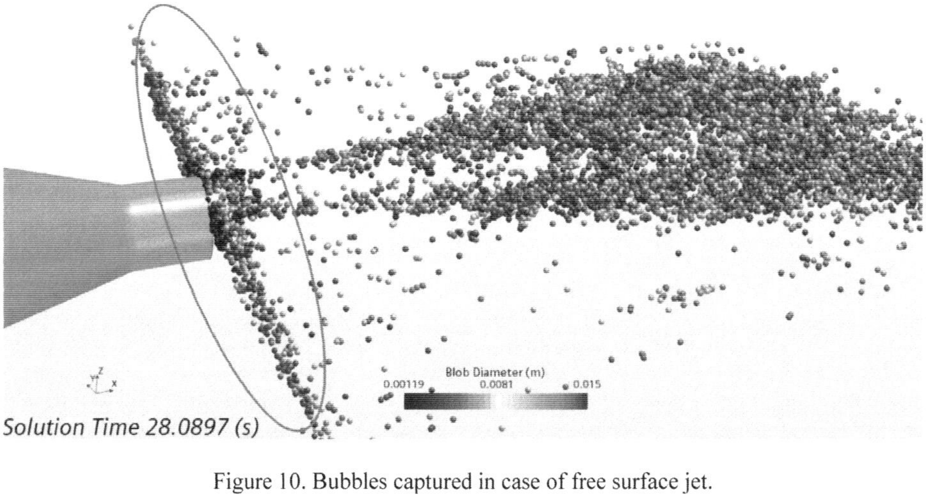

Figure 10. Bubbles captured in case of free surface jet.

Figure 10 shows the blobs of various diameter captured during free surface jet simulation. It can be noted that high number of blobs are captured near to the jet outlet. This is due to the consideration of inlet wall at this location (similar to tank). In practice the jet will be in the wake of the vessel and such effect would not be observed. Accordingly, the computed bubble generation rate is suitably adjusted by discarding the bubbles in the region very near to the jet outlet. Bubble generation rate based on the simulations is shown in figure 11.

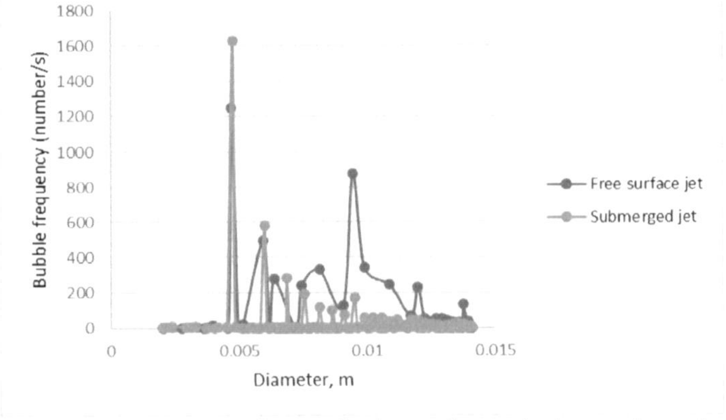

Figure 11. Bubble generation rate for free surface and submerged waterjet

Figure 12a shows the computed PSD of bubble noise considering R_{0_ε}/R_0 equal to 1.5e-5. Referring [9] varied R_{0_ε}/R_0 ratio is discussed. In present case authors consider this ratio would be of the order of 1.5e-6 to 1.5e-5 for the bubbles of 10 – 15mm diameter. And the results of SPL are presented (figure 12b) for these two ratios (assuming constant for all size bubbles). Precise quantification of this ratio would demand further studies and specifically very high computational resources.

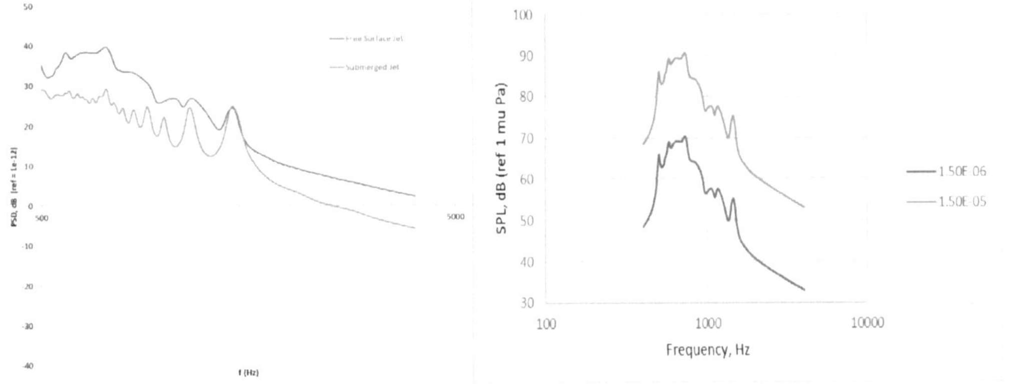

Figure 12. a) PSD of bubble noise for free surface and submerged waterjet (R_{0_ε}/R_0 = 1.5e-5), b) effect of R_{0_ε}/R_0 on SPL (free surface jet case)

5.4 SPL DUE TO WJ NOISE

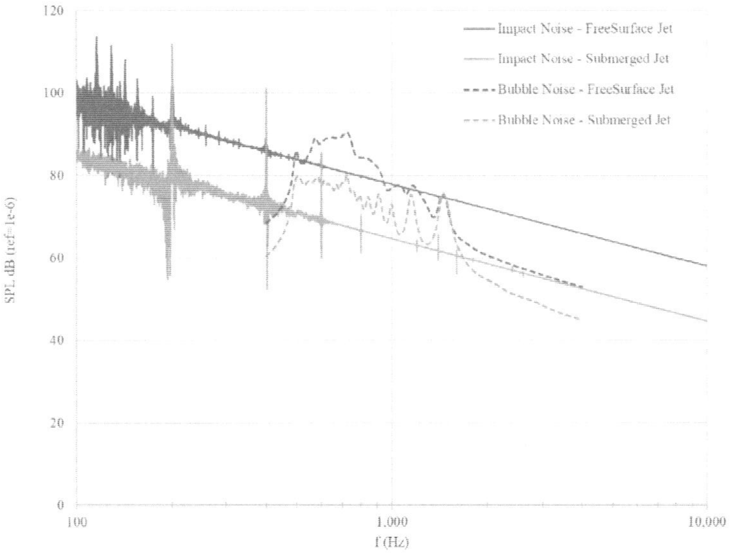

Figure 13. SPL due to WJ efflux

7. CONCLUSIONS

LES based methodology for finding the efflux noise due to waterjet flow is presented. Following are the conclusions from presented methodology

- Compressibility and surface tension effects play important role in acoustic studies of WJ efflux and should be considered in numerical simulations
- Waterjet noise is distinguished into impact and bubble noise.
- Impact noise due to waterjet is approximated based on the oscillating pressure in the fluid domain outside the jet flow region. Basic validation is done in terms of the oscillating pressure as a function of jet velocity. Further validations with the help of experiments would be required.
- Bubble noise
  - It depends upon the mechanism of formation of bubbles.
  - In case of pressurized bubble, the amplitude of vibration of bubble is expected to be high. This case is valid for gas leakage from submerged pipe, air lubrication for ships, etc.
  - In case of waterjet, the bubbles are formed after the impact of jet on water surface. These bubbles are not pressurized, and the vibration amplitude is much less than the pressurized bubbles. The number of bubbles generated depend upon the jet velocity and jet position with reference to free surface.
  - In both cases the bubble generation rate can be reliably estimated with LES based numerical computations. Bubble vibration amplitude is considered suitably, however, requires further investigation for WJ case. Vibration ratio (R_{0_ε}/R_0) of 1.5e-5 is used for bubbles up to 15mm diameter in presented case. In practice, it could be of the order of 3e-4 (e.g. ref [9]) and could lead to higher noise levels compared with impact noise. Such possibilities can be considered at design stage e.g. for stealth vessels.
  - Presented bubble noise model is valid in both cases i.e., pressurised bubble e.g. gas leakage or cavity induced bubble e.g. WJ case
  - The mathematical model considers spherical bubbles, however in reality and in numerical simulations irregular shaped bubbles are generated. Resulting deviations in PSD can be considered acceptable for practical purposes.
- Simulations for submerged WJ indicate the lesser impact and lesser number of bubbles compared to free surface WJ.
- Presented methodology can further be extended for investigating effect of changing jet velocity and jet flow rate to quantify the efflux noise.

8. ACKNOWLEDGEMENTS

The authors express their earnest gratitude to the Indian Register of Shipping, Mumbai for supporting the reported work.

9. REFERENCES

1. FRANZ, G. J., 'Splashes as Sources of Sound in Liquids', *J. Acoust. Soc. Am. 31, 1080–1096,* 1959.

2. PUMPHREY, H., C., CRUM, L. A., JENSEN, L., B., 'Underwater sound produced by individual drop impacts and rainfall', *Acoustical Society of America. Journal, 85(4), 1518-1526*, 1989 (https://doi.org/10.1121/1.397353)

3. PROKHOROV, V., E., CHASHECHKIN, Y., D., 'Sound Generation as a Drop Falls on a Water Surface', *Acoustical Physics Vol. 57, No. 6, pp. 807–818*, 2011.

4. PHILLIPS, S., AGARWAL, A., JORDAN, P., 'The Sound Produced by a Dripping Tap is Driven by Resonant Oscillations of an Entrapped Air Bubble', *Scientific reports 8:9515,* 2018 (DOI:10.1038/s41598-018-27913-0)

5. DOEL, K. V. D, 'Physically-based Models for Liquid Sounds', *ACM Transactions on Applied Perception Volume 2 Issue 4 pp 534–546,* October 2005 (https://doi.org/10.1145/1101530.1101554)

6. MINNAERT, M., 'On musical air-bubbles and the sound of running water', *Philosophical Magazine, 16 (104): 235–248*, 1933 (doi:10.1080/14786443309462277)

7. T. G. LEIGHTON, P. R. WHITE, 'Quantification of undersea gas leaks from carbon capture and storage facilities, from pipelines and from methane seeps, by their acoustic emissions', *Proc. R. Soc. A 2012 468*, 2011 (DOI: 10.1098/rspa.2011.0221)

8. LI, C., NORWOOD, C., WHITE, P., LEIGHTON, T., 'Study of bubble formation dynamics based on associated acoustic radiation', *Intl congress on sound and vibration*, 2017

9. LI, C., WOOD, S., MOORE, S., NGUYEN, B., 'Acoustic Emission of Bubbly Flow and Its Size Distribution Spectrum', *Proceedings of Acoustics 2012 – Fremantle, Australia,* 2012

10. LUO, X., LI, Q., ZHANG, Z., ZHANG, J., 'Research on the underwater noise radiation of high-pressure water jet propulsion', *Ocean Engineering, Volume 219*, 2021, (ISSN 0029-8018, https://doi.org/10.1016/j.oceaneng.2020.108438.)

9. AUTHORS BIOGRAPHY

Sharad Dhavalikar holds the current position of Principal Surveyor at Indian Register of Shipping. He is responsible for research activities in hydrodynamics and Multiphysics studies for ships. His previous experience includes seakeeping assessment and noise predictions for ships, etc.

Ram Kumar Joga holds the current position of Surveyor at Indian Register of Shipping. He is responsible for research activities in CFD studies for ships. His previous experience includes flow investigations for commercial and naval ships, etc.

Praveen Kumar CH holds the current position of Surveyor at Indian Register of Shipping. He is responsible for research activities in numerical hydrodynamics for ships. His previous experience includes seakeeping analysis for ships, etc.

Akula Chatruvedi holds the current position of Surveyor at Indian Register of Shipping. He is responsible for research activities in Multiphysics studies for ships. His previous experience includes underwater and on-board noise predictions for commercial and naval ships, etc.

AN INTRODUCTION TO HYPER LIFT

B Davies and **J Davies,** Dpulsar Limited, New Zealand

SUMMARY

Dpulsar Limited has discovered a concept they designate as "Hyper Lift". Lift explains how seagulls hover, aeroplanes fly, and propellers push huge ships efficiently across oceans. The principles of lift have been precisely defined and its attributes are usually maximised by shapes called foils. Hyper Lift was discovered due to the destructive issues that arise from lift forces within enclosed contra-rotating jets. Contra-rotating impellers can produce enormous energy that, if unmanaged, causes catastrophic failure to the device. The Dpulsar team proposed that this energy may be harnessed to create an extremely powerful high mass pump. The resulting Micro-Pulse jet utilises contra-rotating dynamics to enable Hyper Lift; substantial lift on the upstream impeller blades which enhances the effect of the foils. Consequently, acceleration of the main body of water within the pump is considerably increased. Hyper Lift offers great potential for higher thrust and superior efficiencies in fluid propulsion.

NOMENCLATURE

Thrust	The force used to discuss propulsive ability of a system operating on a fluid. This force is comprised of two components: mass (m) in kg and change in velocity (Δv) in ms^{-2}. The proportion of m to Δv significantly alters thrust properties. The magnitude of thrust is the product of these two components.
Fluid Continuity	The principle that incompressible fluids maintain an inverse proportion between the cross-sectional area of the fluid and the change in velocity. When water is forced through a halved cross-sectional area, this corresponds with a doubling of the velocity of the fluid. Conversely, when water is accelerated to double its previous speed, it occupies half of the original area.
Propulsive Efficiency	A measure of the portion of the available energy in the system that is usefully applied in propelling the vessel, relative to the total energy of the jet stream.
High Pressure System	A marine propulsive system designed to achieve water acceleration by generating high internal water pressure and directing that water through a reducing nozzle. This causes the fluid to accelerate.
Low Pressure Contra-rotating System	A marine propulsive system designed to accelerate water over a pair of enclosed contra-rotating impellers at the lowest possible internal water pressure.
Stalled State	A situation that may occur in enclosed contra-rotating systems. In this state, both impellers' ability to generate lift is significantly impeded. Each impeller has stalled. Jet propulsion is therefore interrupted. When enduring, this state results in hydraulic lock-up and system failure.
Hyper Lift	The state achieved by artificially enhancing fluid acceleration over the lifting surface of a foil through utilising a boundary layer flow. The resulting lifting force is greatly increased.

1. INTRODUCTION

The advantages of contra-rotating propellers as a propulsion system are well understood. However, applying the principle to enclosed waterjets produced seemingly insurmountable problems. While attempting to solve these various issues, the Dpulsar team unlocked a phenomenon they believe takes efficient propulsion to a new level: Hyper Lift.

The Dpulsar contra-rotating jet was configured to accelerate large masses of water with the lowest possible back-pressure. Even with this highly efficient formulation, the issue of enclosed contra-rotation was challenging. The contradictory forces between the contra-rotating impellers resulted in multiple severe failures. Hyper Lift proved to be the solution. This enhanced state allows rotating blades to produce vastly superior lifting characteristics.

Hyper Lift has the possibility to transform propulsion. The marine industry faces several significant issues in the modern era. Efficiency is a critical factor in the transition away from legacy fuels. Hyper Lift presents the opportunity for unlocking superefficient propulsion. Furthermore, the phenomenon has prospects for greater performance in additional applications: pumping, hydroelectric power generation, and even powered flight.

2. THE MICRO-PULSE PLATFORM

The Micro-Pulse jet was designed to generate and utilize Hyper Lift. It is a coaxially driven high mass pump that turns the destructive energy of stalled contra-rotating impellers into highly efficient propulsion. This is achieved by converting the energy created by stalled impellers into manageable pulses of pressure. In turn, this drives a boundary layer of water along an alternative flow-path. Through this, Hyper Lift is created and maintained on the upstream impeller blade surfaces. Once this state is achieved, the previously stalled system begins to operate as an extremely efficient waterjet.

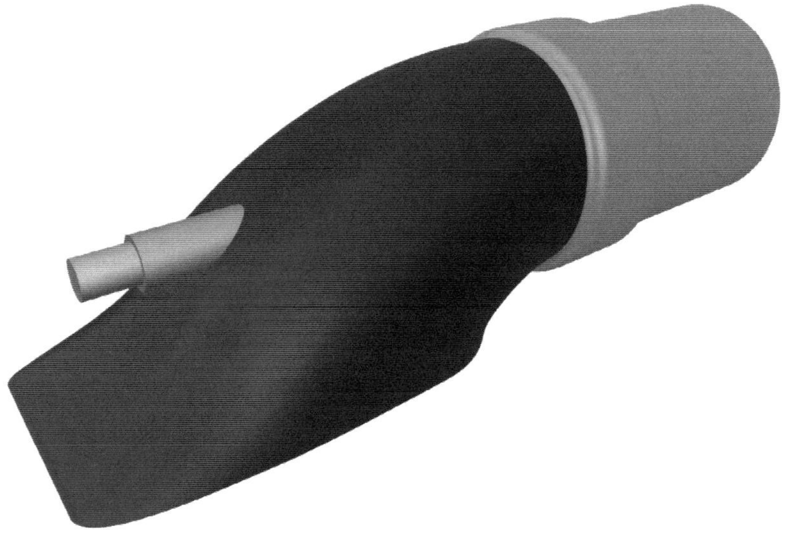

Figure 1. Micro-Pulse contra-rotating jet unit

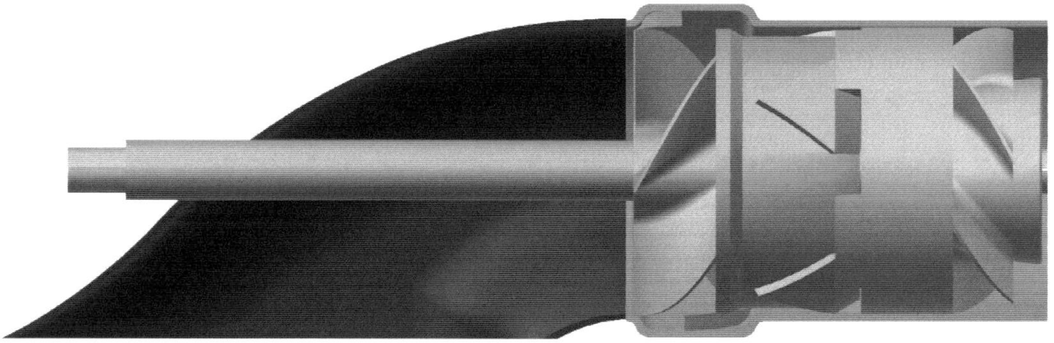

Figure 2. Cutaway profile of the Micro-Pulse jet

The Micro-Pulse system deviates from typical marine propulsors in key areas. Therefore, understanding the different hydrodynamic principles in the various water-jet propulsion devices is necessary at this juncture.

3. BASIC PRESSURE JETS

Axial flow pressure jets use the principle of accelerated water to produce thrust. They use a pressure inducing impeller, capable of developing high pressure. The pressurized fluid accelerates through a reducing nozzle section. The resulting force produces an equal and opposite reaction, sufficient to drive the vessel.

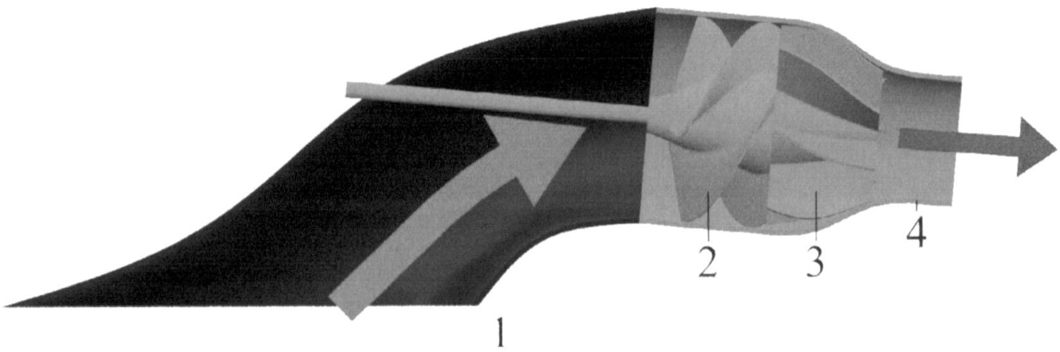

Figure 3. Cutaway profile of an axial flow pressure jet

A basic high-speed pressure jet contains four key components which are displayed in Figure 3. They are the water intake (1), the impeller (2), the stator set (3), and the nozzle (4). The rotation of an impeller draws water from outside the craft via the intake. This is depicted by the blue arrow in Figure 3. The impeller is designed to maintain high water pressure. Ingested water is forced downstream by the impeller's rotational motion. It is directed through a set of straightening stators to reduce radial energy losses. The water continues through a reducing nozzle before exiting into the atmosphere. This is depicted by the red arrow in Figure 3. The outlet cross-sectional area relative to the impeller blade cross-sectional area dictates the change in velocity. In high-speed axial flow pumps this reduction ratio is typically 1:0.27. Through fluid continuity, this indicates a velocity change of almost four times. In mixed flow and centrifugal jets the reduction ratio is typically even larger, producing even greater change in fluid velocities.

This typifies a high-pressure system. In accordance with Bernoulli's principle, the highest pressure is directly downstream from the impeller, and the highest velocity is where the pressure is at its lowest, just as the fluid exits the nozzle. These devices require tight tolerances to maintain high pressure head.

4. BASIC CONTRA-ROTATION

In the early 1990s, Dpulsar's lead developer Barry Davies was intrigued by the potential of applying opposite-rotating impellers to a pump operating above the waterline. Such a contra-rotating system replaces the stator set shown in Figure 3 with an opposite pitched spinning impeller. This avoids the need for fixed straightening vanes; contra-rotation promised to reclaim energy, rather than simply reduce losses. The downstream impeller, with its opposite pitch and contra-rotating blades, recovers the radial energy imparted by the upstream impeller. The result is axial flow at the outlet. However, when the unit was designed to generate high pressure, the fluid acceleration was occurring in the nozzle, which was exactly the same as an axial-flow jet. The predicted gains from using a contra-rotating impeller over a stator set did not eventuate; the pump performed identically to a pressure jet.

4.1 HIGH MASS CONTRA-ROTATION JETS

The breakthrough came when it was realized that contra-rotating thrust could be generated by moving a larger mass at lower pressure, instead of highly pressurizing a smaller mass of water like a traditional jet. The two impellers were then designed to operate on a larger mass of water per revolution, accelerating the fluid across the blades. The outlet cross-sectional area was altered too; it was opened to reduce backpressure and minimise restriction at the outlet, as these variables would lock up the unit. This was precisely balanced against being large enough that it allowed ventilation, a state where air being drawn into the outlet de-primes the device. In this new configuration, the design conformed to Froude efficiency, achieving higher propulsive efficiency at low, medium, and high speed [1]. As a result, the thrusting characteristics and propulsive efficiencies were more comparable to propeller systems. This unit furthermore achieved similar speeds to a high-speed axial flow jet.

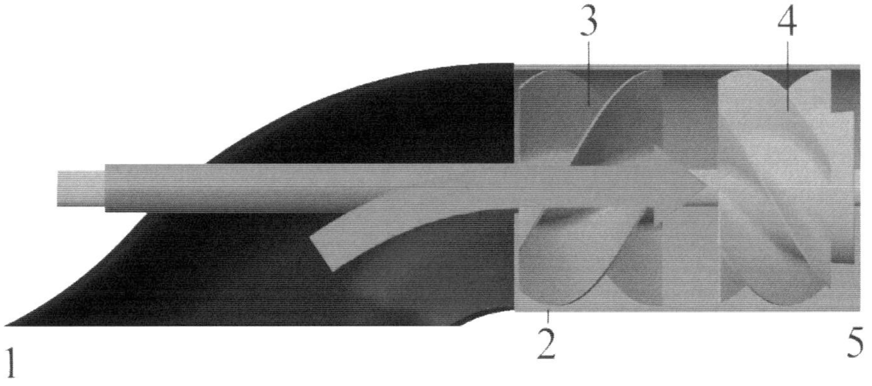

Figure 4. Cutaway profile of a contra-rotating jet configured for high mass

The high mass contra-rotating jet has five main components, as shown in Figure 4. They are a fluid intake (1), a pump housing (2), an upstream impeller (3), a downstream impeller (4), and an outlet (5). Water is drawn from the intake by the rotating impellers, as shown by the blue arrow in Figure 4. The impellers accelerate the water, and the water is deposited out of the outlet.

The high mass contra-rotating jet is distinct from a traditional jet in a few key areas. The impeller blade pitch is coarse, differing from a waterjet by more than 30°. In this design the impellers are pitched to around 45° at the blade tips. A 190mm diameter high mass contra-rotating unit is geared to around half of engine RPM, running far slower than a traditional pressure jet. A high-speed jet accelerates water in a nozzle section, and accordingly uses a significant cross-sectional area reduction in the nozzle. In contrast, contra-rotating pumps accelerate water over the impellers and use an outlet area reduction proportion ratio of 1:0.55. According to fluid continuity, this corresponds to a change in velocity for the fluid of approximately double. These differences are indicative of the fundamental difference in how the two systems function.

This pump is configured so that the mass component per each impeller revolution is set at a maximum. Consequentially, the pressure-driven need for tight tolerances is relaxed. The high mass device is more resilient to internal damage. However, the contra-rotating pump acts more like a ramjet in stationary situations. It requires forward motion to generate sufficient intake pressure to be able to apply full power. High pressure waterjets are able to have full power applied in stationary situations which for some applications is a substantial advantage.

The multitude of variables, rotational speeds, shapes, and pitches made development very challenging. Additionally, there proved a fundamental issue to successfully operating the system.

4.2 THE CHALLENGE OF ENCLOSED CONTRA-ROTATING IMPELLERS

Enclosed high mass contra-rotation jets had proven to be difficult as there was a significant hurdle; the pressure field of the downstream impeller was a source of competition for the upstream impeller. This was due to the enclosed nature of the pump. Without overcoming this issue, the two impellers' effective pressure fields would decay, and each impeller would stall. With this hydraulic lock-up, an immense amount of energy would be generated. This caused catastrophic damage to the unit. After intensive development, the successful relative impeller speeds were identified to prevent this damaging state.

Dpulsar was established to investigate whether the energy from hydraulic lock-up could be harnessed. Instead of preventing the competition between impellers, the Dpulsar team hypothesized that it could be managed to intentionally allow a large build-up of energy; once redirected, this could be used for a positive purpose. As the system was based on lift-generating impellers, this was the first point of investigation.

5. THE BASICS OF LIFT

Modern understanding of the propeller is widely attributed to the Wright brothers. In the early 1900s, they discovered the theoretical basis of lift and sustained flight. This breakthrough further explained how propellers achieved far more propulsion than simple screws. The cross-section of a modern wing or blade is asymmetric, with the raised section referred to as a foil. The fluid accelerates over the foiled side faster, relative to the non-foiled side. The accelerated fluid has an associated drop in pressure. This pressure differential results in lifting force, as displayed in Figure 5.

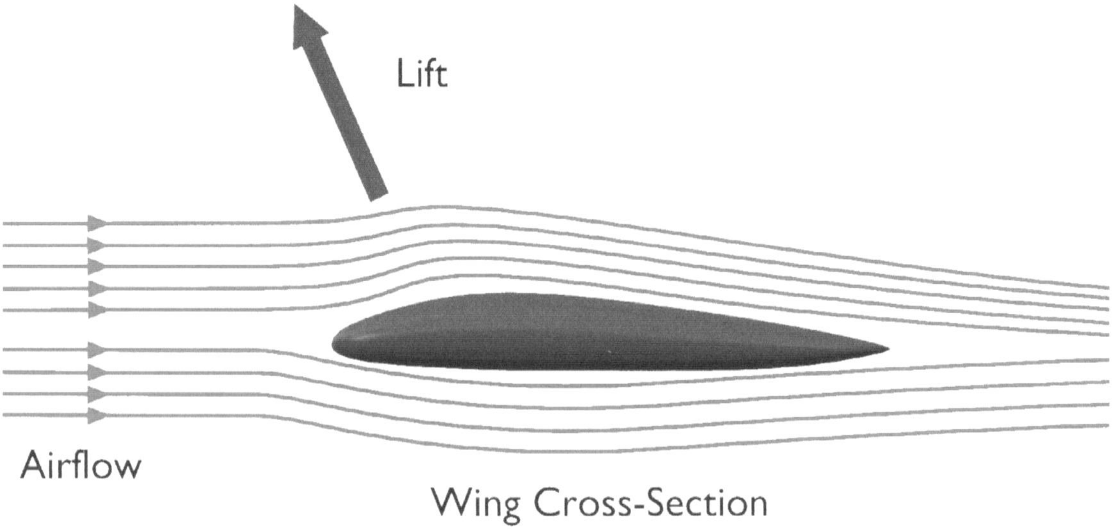

Figure 5. The effect of a foiled surface on fluid flow

Modern propellers operate functionally like a set of wings fixed to a central point of rotation. For a propeller, each rotation causes the fluid to accelerate rearwards at an increased velocity. This produces an equal and opposite reaction, the resultant thrust, being sufficient to propel a vessel forward. As they operate within the water, they do not need to straighten flow to maintain thrust. The various designs of foils enhance a propeller's efficiency for specific applications.

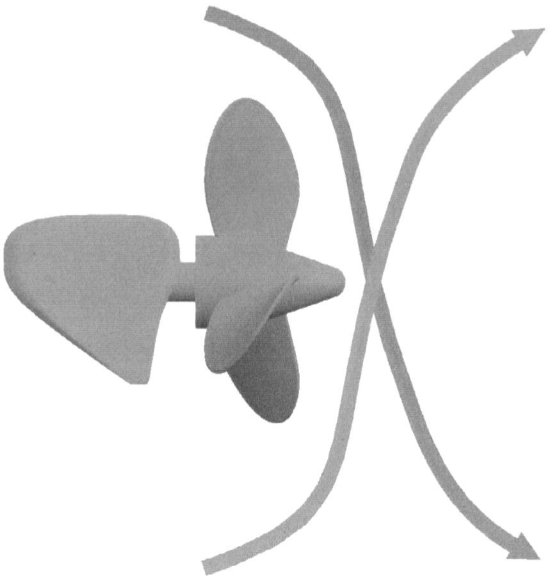

Figure 6. Profile of a propeller operating in water

Propellers have excellent overall propulsive efficiencies. They can operate effectively over a wide range of weather and water conditions. Lift is the key to propeller propulsion. Fluid is accelerated across the blades' surface while the foil adds to the lift vector to give a higher advance coefficient. The lowest water pressure is directly downstream from the propeller, where the water is at a maximum velocity. This is axiomatically different from a waterjet.

5.1 LIFT IN A HIGH MASS CONTRA-ROTATING JET

As previously mentioned, effective and enclosed contra-rotation requires a high mass configuration. Such systems follow the principles of lifting propellers, operating at the lowest possible back-pressure. This is distinct from high pressure waterjets, which require significantly raised pressure to induce a velocity change in the fluid. In this high mass pump, two coarse and opposite pitched impellers enable the mass component of thrust to be maximized and the velocity change to be kept at a minimum.

5.2 STALL

This unit posed a unique problem. Unless the impeller rotational ratios were set to a precise range, the devices would hydraulically seize. Enclosing the impeller blade tips with a pump housing's solid wall forced the two impellers to compete directly with each other. The low-pressure lifting face of each downstream blade appeared to interfere with the high-pressure face of each upstream blade. This led to the impellers losing their ability to generate lift. Effective water flow through the jet was obstructed. When this happened enormous energy was produced, sufficient to lose propulsion and destroy the device. In this situation, the impellers were deemed to have stalled, as they no longer resulted in significant thrust.

The Dpulsar team postulated that the energy from the stall could be harnessed to give a propulsive advantage. They proposed that a secondary flow-path within the pump could be established using the energy from the stall. This secondary path would induce a divergent, accelerating boundary layer, which would need to generate enough lift on the upstream impeller blades to overcome the stall. If a boundary layer of water could be established, then the direction of flow over the impeller blade tips could also be modified. This was a difficult concept; such a hypothesis required a disadvantage and an advantage to simultaneously operate to make the device work.

This was successfully achieved through the Micro-Pulse technology. The energy was redirected back into the system, enhancing impeller efficiency through artificially increasing lift. Hyper Lift was now a reality.

6. MICRO-PULSE

The resulting invention is the Micro-Pulse jet unit. This high mass, contra-rotating pump harnesses the destructive energy of stalled contra-rotation. Through converting this energy into manageable pulses, the Micro-Pulse system creates a secondary flow-path. This boundary pathway redirects water over the upstream impeller's blade tips to improve its lifting ability. Hyper Lift is the result, producing vastly superior thrust through all speed ranges.

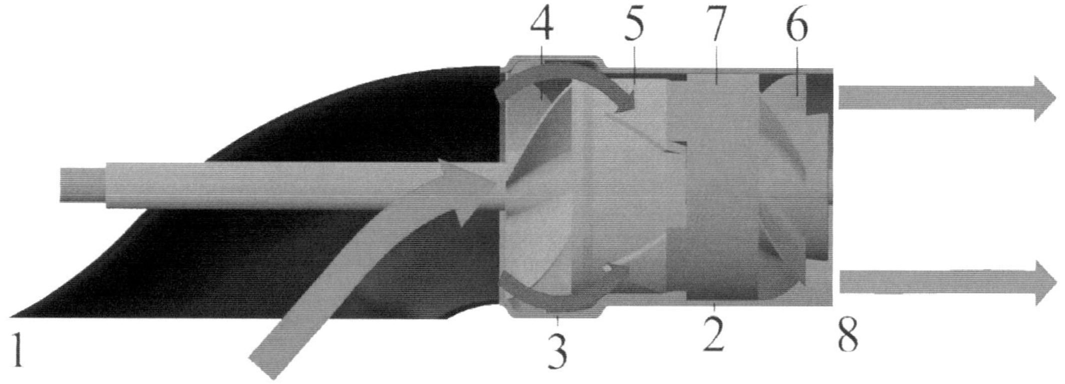

Figure 7. Cutaway profile of the high mass contra-rotating Micro-Pulse jet

The Micro-Pulse system has eight essential components, as shown in Figure 7. They are a fluid intake (1), a pump housing (2), a boundary flow bypass (3), an upstream impeller (4), an upstream impeller shroud (5), a downstream impeller (6), a downstream impeller shroud (7), and an outlet (8). Water is drawn through the fluid intake by the rotating impellers, and deposited out of the outlet as straightened wash, shown by the blue arrows in Figure 7. The unit also contains a secondary pathway, which accelerates a boundary layer of water over the upstream portion of the upstream impeller blade tips. This is shown by the red arrows in Figure 7. The bypass flow-path deposits its accelerated water into the downstream impeller.

Both impellers have shrouds containing a series of channels and gates. These channels and gates phase in and out of sequence as the impellers contra-rotate.

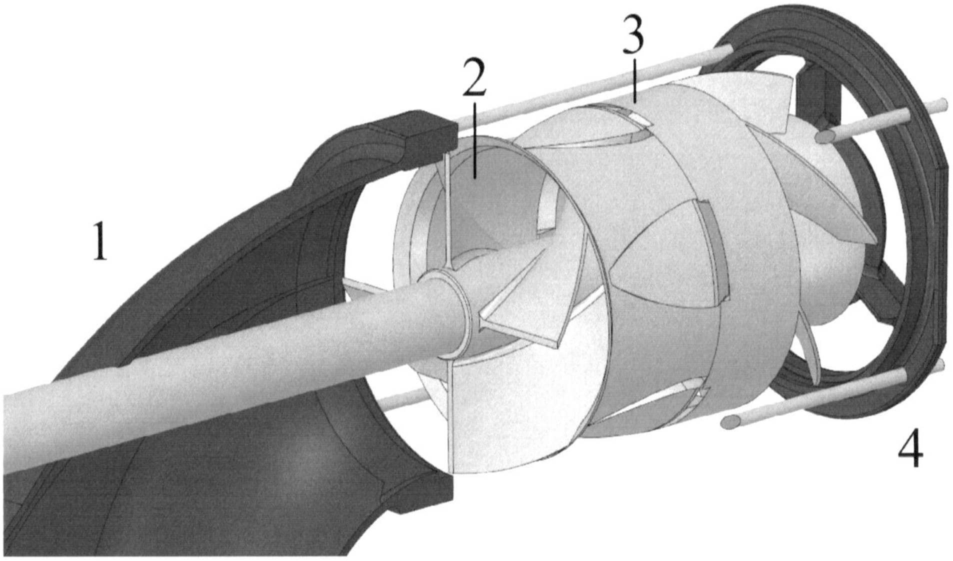

Figure 8. Cutaway representation of the Micro-Pulse system in a closed gate state

Figure 8 depicts the Micro-Pulse unit with four relevant components: the fluid intake (1), the shrouded upstream impeller (2), the shrouded downstream impeller (3), and the rear support structure (4). The pump housing and boundary flow bypass are omitted for clarity. Figure 8 shows the jet in the rotational stage where the gates are closed. This causes the system to experience a momentary stall. When in a stalling alignment, micro-pulses of extremely low pressure are generated. The prototype Micro-Pulse unit generates ten of these pulses per each complete rotation of the impellers.

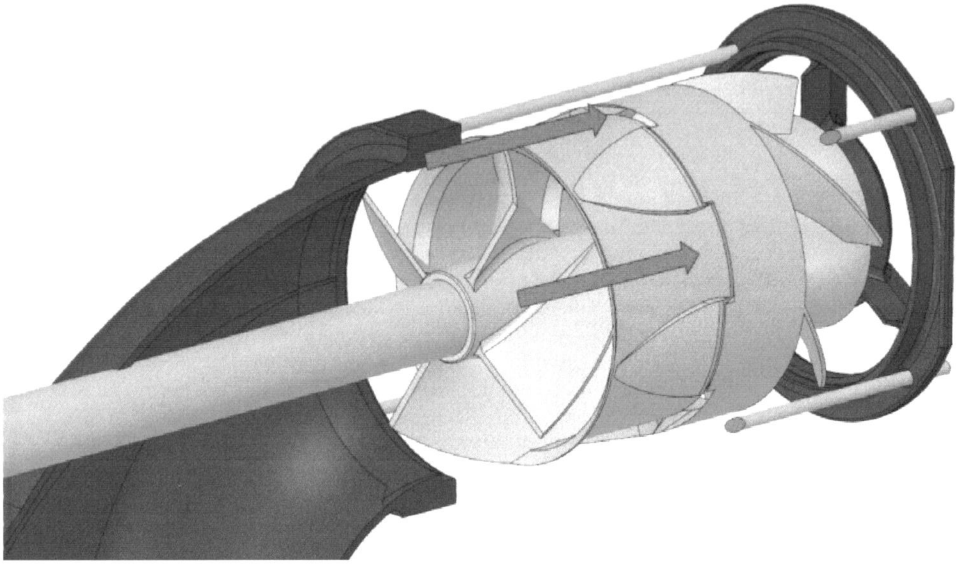

Figure 9. Cutaway representation of the Micro-Pulse system in an open gate state

Figure 9 shows the same components as Figure 8. However, the system is depicted with its gates open. The resulting boundary flow-path is represented using blue arrows.

During unit rotation, the gates progressively open, creating a connection along the channels to the area surrounding the upstream impeller's blade tips. Consequently the two pressure zones are momentarily connected. The pressure differential causes an axial acceleration of a thin boundary layer of water from the outer edges of the upstream impeller blades down the isolated channels. The main body of water, unlike the boundary layer, is rotated helically by the motion of the blades. The divergent and axially accelerating boundary layer develops the lift profile of the upstream impeller. This is the phenomenon Dpulsar describes as Hyper Lift. The enhanced lift is sufficient to un-stall the unit and creates improved change in velocity through the main body of the pump. The downstream impeller is now able to recover the radial energy of the water, to produce the required axial wash. Higher thrust and greater efficiencies are achieved as a result. The Micro-Pulse pump is fully enclosed, having all the advantages of a waterjet as well as those of a propeller.

Figure 10. Wash profile of a Micro-Pulse jet

Two flow paths exist within the Micro-Pulse system and are evident in the unit's wash, shown in Figure 10. Like high mass pumps, the main body of water is deposited out of the outlet, rather than accelerated through it. The outlet maintains just enough back-pressure to ensure jet priming and to avoid ventilation. The secondary flow-path operates entirely differently; it retains rotational energy imparted by the downstream impeller. This thin layer is released through the outlet along with the main body of now axially flowing water. Figure 10 shows the low-pressure axial flow as wash, alongside the rotating boundary layer spray. This diffuse spray has intentionally not been suppressed to display the occurrence.

The Micro-Pulse pump shares many operating similarities to a simple contra-rotating jet. It utilises coarse-pitched impellers operating primarily on lifting principles. The system is relatively slow rotationally; at 190mm unit diameter, the pump runs geared at half of engine RPM. Like a high mass contra-rotating jet, the device is designed for minimum back-pressure. Many configurations were trialled to evaluate the effective range of the outlet. The outlet size was required to match the associated acceleration of water through the unit due to fluid continuity. Too large an outlet would allow the system to ventilate by drawing air through the outlet and preventing priming of the pump. Too small an outlet would result in pressurisation of the device, which negatively influenced its effectiveness. The successful outlet size results in a reduction ratio of 1:0.31. This correlates with a triple acceleration of water through the jet. For the same unit RPM, this surpasses the double acceleration achieved in contra-rotation without Hyper Lift.

The Micro-Pulse jet operates with superior properties to alternative propulsors. The attributes of Hyper Lift overcome the previous shortcomings of high mass contra-rotating pumps; the Micro-Pulse unit can be run at full input speed in a stationary position. Due to its nature as a high mass system, the Micro-Pulse invention is scalable to alternative pump diameters. It is mechanically simple through using the existing engineering necessary for contra-rotation. Hyper Lift requires no outside energy inputs beyond the main drive, allowing the driven Micro-Pulse unit to be self-sustaining. The diffuse axial plume is equally suited for both surface and subsurface use, with minimal environmental impact. The system is fully enclosed and presents no obstacles protruding beyond the vessel hull-form. It is thus streamlined and safer for the vessel and crew. The steering apparatus is located in the wash which gives superior control and manoeuvrability. The jet offers versatility, resilience, safety, and powerful, efficient performance. It creates more thrust with less energy required. It is not dependent on any particular power plant and is as easily coupled to an electric motor as to an internal combustion engine. This technology has great potential in reducing carbon footprint in the marine industry as well as helping to transition to alternative power systems, such as electric motors.

The benefits and performance of the Micro-Pulse system are entirely dependent on Hyper Lift. Without it, the impeller competition is a major obstacle to system integrity. However, mitigating the impeller competition also affects system performance. Hyper Lift turns the competitive destruction into an advantage.

7. HYPER LIFT

Hyper Lift is the result of using a boundary layer flow-path to artificially augment the effect of a lifting foil. As shown by the lift diagram in Figure 5, the accelerating fluid over the foiling face is essential to lift, and therefore propulsion. Hyper Lift is achieved by the amplified acceleration over the foiled face; this increases the pressure differential over the blade, and results in greater lift. The dynamic of Hyper Lift has exciting potential. Using Hyper Lift, many existing lift-based systems could be enhanced beyond what traditional development has achieved.

In water, the Hyper Lift phenomenon creates efficient, powerful, and safe thrust above and below the waterline. The potential for Hyper Lift in current marine applications is shown by the Micro-Pulse jet. Hyper Lift could already be used to improve stationary low-head pumping and hydro turbine power generation. As it is a new fluid dynamics principle, the Dpulsar team anticipates that Hyper Lift could offer the same gains in air as it does in water. This suggests possible advances in aeronautics.

The Dpulsar team believes that Hyper Lift is the critical discovery behind the Micro-Pulse invention. It is early in the development of the system and investigations of the underlying phenomenon. However, Hyper Lift's enhancement promises significant opportunity. Hyper Lift offers substantial contributions to more sustainable and efficient propulsion at a time where large departures from the norm are becoming essential.

8. CONCLUSIONS

Through the Dpulsar team's investigation into high pressure and high mass contra-rotating jets, a new phenomenon has been discovered. By producing increased thrust through Hyper Lift, the Dpulsar team was able to overcome the issues present in enclosed high mass contra-rotating jets. Furthermore, Hyper Lift itself is an extremely important breakthrough in fluid movement. The importance of being able to artificially induce lift on an impeller blade surface, without the need to modify shapes or rotational speeds, cannot be understated. Hyper Lift allows the lifting characteristics of an impeller to be magnified considerably, even at comparatively low tip speeds and without highly foiled surfaces.

This offers significant steps forward in lift-generating devices. It has the potential to advance propulsion to the next stage. The timeliness of this discovery is apparent: there is significant pressure to transform maritime practises. Hyper Lift systems such as the Micro-Pulse jet are more efficient and can be easily coupled to electric motors or internal combustion engines. Hyper Lift unlocks efficient, scalable, superior performance in a variety of applications: hydro and air propulsion, hydro pumping and power generation.

9. ACKNOWLEDGEMENTS

The Dpulsar team are grateful to C-Fury Limited for their independent systems testing in the UK. Their trials using a Dpulsar contra-rotating jet with a prototype belt drive to evaluate different power plants are appreciated.

10. REFERENCES

1. JOBSON, D. and WALSHAW, A., 'Fluid Momentum and Thrust by Reaction', *Mechanics of Fluids,* 1979.

11. AUTHORS BIOGRAPHY

Barry Davies holds the current position of lead developer at Dpulsar Limited. He is responsible for inventing, research, development, testing, and technical writing for patents. His previous experience includes 45 years of work with jet boats and inventing viable high mass contra-rotating jet units under Contrapel Limited.

John Davies holds the current position of assistant developer at Dpulsar Limited. He is responsible for co-inventing, research, development, testing, 3D modelling, and computer aided design. His previous experience includes a Bachelor of Science in Mathematics and Computer Science from the University of Canterbury.

NUMERICAL SIMULATION AND ANALYSIS OF THE PRESSURE FLUCTUATION OF FOUR WATERJETS IN V-TYPE SHIP

Y Dai, J Wang and **J Liu**, Science and Technology on Water Jet Propulsion Laboratory, China

SUMMARY

The noise and vibration induced by pressure fluctuation is one of the major issues that threatens the waterjet-propelled ship comfortability. The self-propulsion simulations of four waterjets which are installed at the V-type Vessel's Stern are carried out by solving the unsteady RANS equations. The analysis on pressure fluctuation characteristic for non-cavitation conditions are presented on the inner surface of the waterjets 5 locations along the axial direction, which are duct outlet, impeller inlet and outlet sections, impeller tip, stator outlet. This paper also focuses on comparing the pressure fluctuation characteristic between the four waterjets which are installed at the V-type Vessel's Stern and the flush type of intake waterjet. This research could be useful for structural design of the waterjet propulsion ship.

1. INTRODUCTION

Waterjet propulsion is widely used in high-speed marine vessels with several advantages comparing with propeller propulsion, waterjet propulsion has no appendage, low resistance, and superior cavitation resistance under high-speed conditions, the waterjet has been widely used in high-performance ships. The pressure fluctuation in the pump is one of the important reasons for the vibration and noise of the impeller. Rickard E Bensow et al. (2017). Analysis the effect of impeller tip clearance on hull pressure pulses. The pressure fluctuation is mainly caused by the non-uniform distribution of the flow velocity in the inlet duct, the vortex generated by the flow, and the interference between the pump impeller and the stator.

With the rapid increase of the tonnage and main engine power of water-jet propulsion ships, the traditional dual-water-jet propulsion method can no longer meet the requirements of ship rapidity. Therefore, three, four or more waterjets are used on some high-speed and high-power water-jet propulsion ships. The arrangement of multiple impellers is different from that of single or two impellers. The inlet flow field of two adjacent impellers will interact with each other, and the tail line of v-shaped stern will affect the height of the axis. All these factors will change the operation characteristics of impellers. In recent years, with the development of CFD numerical simulation technology, the hydrodynamic performance of the interaction between waterjets can be estimated effectively. Tomohiro Takai(2011), Kandasamy et al.(2013) conducted a series of studies on self-propulsion and numerical simulation of surface ships based on CFDShip-IOWA and using overlapping grids and carried out confirmation and verification studies on self-propulsion simulation of high-speed axial jet propulsion Ship models using URANS method. Zobeiri et al. (2006) conducted numerical simulation and analysis on stator and impeller interference in the pump.

In this paper, the research of the model scale v-shaped ship equipped with four waterjets is carried out. There is obvious deadrise angle at the aft of the hull, intermediate hull draft is deeper, the side draft gradually reduced, as shown in Fig. 1. The hull and waterjet geometries were at the model scale. Four waterjet systems were installed in the hull, the principal dimensions of the hull and waterjet are shown in Table 1. The axis of the lateral waterjet slightly higher than the inside waterjet, the initial waterline slightly higher than the waterjet hydraulic axis.

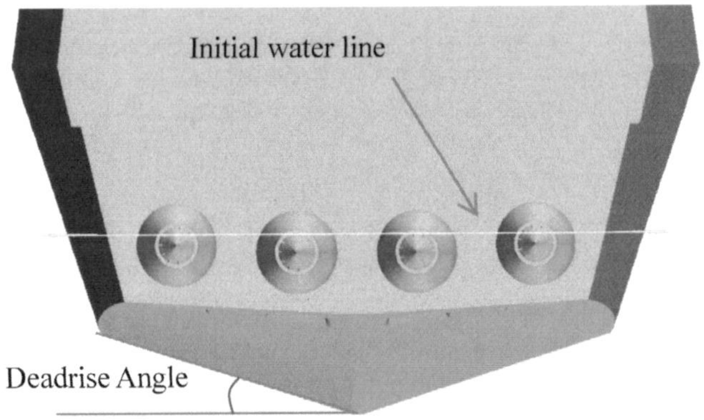

Fig. 1 Four waterjet installed in the stern of V-shaped ship

Table 1. Principal dimension of the hull and waterjet

	Hull		Waterjet
LOA(m)	6.98	Diameter(mm)	103.6
B(m)	0.8133	Nozzle diameter (mm)	50
T	0.3867	Impeller blades	6
∇ (m3)	0.579	Stator blades	11

The inflow characteristics of the inlet duct of the inner and outer waterjets were obtained through three-dimensional steady numerical simulation, and the unsteady numerical simulation was carried out to explore the characteristics of the inlet and outlet of the impeller, the clearance between the top of the impeller and the shell, and the circumferential pressure fluctuation of the shell at the five positions of the guide blade nozzle, so as to provide a reference for the noise control of the waterjet.

2. NUMERICAL METHODS

Unsteady RANS model

The SST k-ω model was applied to simulate the three-dimensional unsteady viscous incompressible flow in the waterjet pump in Starccm+. The basic N-S equation is introduced into the Boussinesq hypothesis, and the obtained flow control equation is the continuity equation in time-homogeneous form and the N-S equation (Eqs. 1~2).

$$\frac{\partial(\rho\overline{u_i})}{\partial x_i}=0 \qquad (1)$$

$$\frac{\partial(\rho\overline{u_i u_j})}{\partial x_i}=-\frac{\partial p}{\partial x_i}+\frac{\partial}{\partial x_j}[\mu(\frac{\partial(\rho\overline{u_i})}{\partial x_j}-\rho\overline{u_i u_j})]+\rho F_i \qquad (2)$$

Fi is the volume force, P is the pressure acting on the fluid, ρ is the density of the medium; ν is the kinematic viscosity. The standard SST model is used to close the RANS equations by introducing a two-equation model of the dissipation rate of turbulent kinetic energy.

VOF model

Ship computational fluid dynamics usually involves the simulation of free surface. For the numerical simulation of self-propulsion of a waterjet ship, the problem of free surface needs to be taken into account because the water flow injected directly into the air. In fact, the treatment of free surface problem is to track the motion interface between two media. The existing numerical solutions to the motion interface Tracking problem mainly include Volume of Fluid, Level set, and Front Tracking Method. The VOF model was proposed by Hirt and Nichols in 1981. This method can deal with steady-state and transient problems with interface between air and liquid. The VOF model solves the same set of momentum equations for the insoluble fluids throughout the computational domain and tracks the volume fraction of each fluid to simulate the multiphase flow. Based on the VOF method of ship two-phase flow problem, unit interface flux interpolation format has a geometric reconstruction (Geo-reconstruct) format, donor acceptor (donor-acceptor) format, modified high resolution interface to capture format (HRIC), etc. The geometric reconstruction scheme assumes that the interface of the two-phase fluid in each cell is a polyline with a linear slope, and the linear slope is used to calculate the flow of the fluid through the cell interface. In the donor-recipient format, the direction of the interface of different phases is horizontal or vertical. According to the direction and movement of the interface, the pass value is obtained by means of upwind, downwind or a combination of the two interface interpolation formats. The modified HRIC format adopts the donor-recipient method, including the combination of windward difference and leeward difference. The geometric reconstruction format has the highest accuracy and the largest amount of calculation, followed by the correction of HRIC format and donor-acceptor format.

Mesh and boundary conditions

Considering the ship's direct motion and the strict symmetry of the hull, only the semi-ship model is numerically simulated to reduce the calculation, and the boundary conditions of symmetrical plane are used in the middle longitudinal section. In order to ensure the steady flow at the inlet and outlet of the computing domain, the inlet and outlet of the computing domain are 2 times and 5 times of the captain, and the bottom and top of the computing domain are 2.5 times and 1.5 times of the captain, respectively. External and internal watershed hydrology of flow channel by hull Trimmed grids, grid, the boundary layer with three prisms to import port lip and rounded corner's location has a good joint effect, lifting Angle of mesh quality. At the same time, the bow, tail, waterplane and waterjet inlet of the main body and body are encrypted, and the ship's traveling wave encryption area is set to gradually encrypt in the direction close to the hull, which is conducive to capturing the pressure distribution of the bow and stern. A honeycomb polyhedral mesh is used in the stator region. The height of the first layer of grid is selected according to the range of wall y+ between 30 and 60. In order to capture velocity and pressure in the boundary layer of the inlet duct, 20 layers of the boundary layer are set at a growth rate of 1.1, and local mesh density is increased in the area near the inlet duct, as shown in Fig. 2.The total number of grid nodes in the whole computing domain is about 11 million, including 7 million in the basin and inlet duct, and 2 million in the inner and outer impeller domain, as shown in Fig. 2-4.

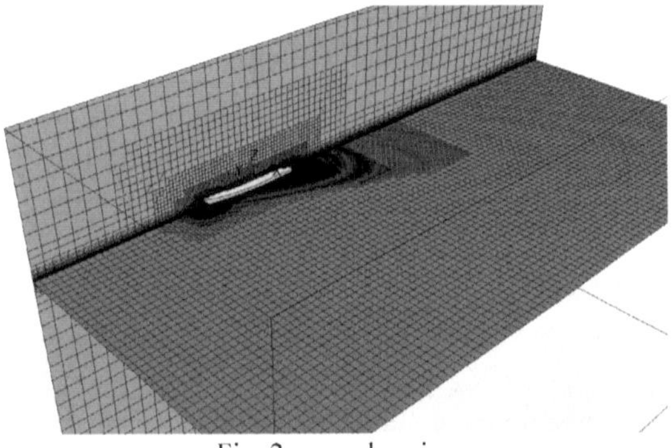

Fig. 2 general region

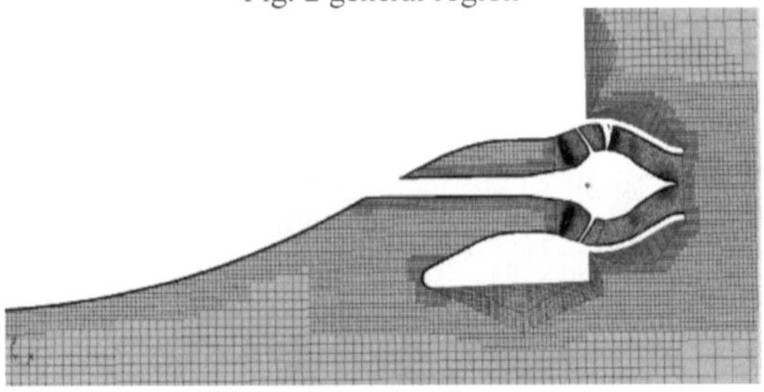

Fig. 3 meshgrids of waterjet

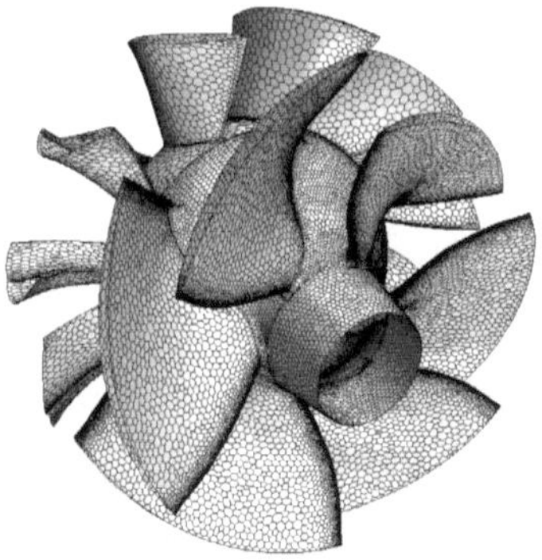

Fig. 4 meshgrids of waterjet pump

VOF (Volume of Fluid) model is used to simulate the free surface, Correction HRIC (High Resolution Interpolate Correction) format is used as VOF interface interpolation format. The inlet boundary of the calculation domain is velocity inlet, the outlet boundary condition is pressure outlet, the lateral and middle longitudinal plane are symmetry boundary condition, and the top and bottom are velocity inlet condition. During the calculation process, the ship's speed and attitude remain unchanged, and the MRF model is used to set the rotating coordinate system to simulate the rotation of the impeller. The SIMPLE method is used for pressure-velocity coupling iteration. The central difference scheme is used for the pressure equation, and the second-order upwind scheme is used for the momentum equation, the turbulent kinetic energy equation and the dissipation rate equation.

In this paper, the design condition of the impeller is 1900r/min. The unsteady numerical simulation and steady simulation results are taken as the initial value. If the sampling frequency is too low, the pressure fluctuation amplitude will be underestimated. The smaller the unsteady simulation time step is, the higher the sampling frequency is calculated, and the pressure fluctuation amplitude is the more accurate. In this paper, the unsteady simulation time step is 1/(1900/60)/360=0.000121s. The total calculation time is 0.14s.

3. GENERAL FLOW BEHAVIOUR

The outlet of the inlet duct of impeller is the inlet of waterjet propulsion pump. The non-uniform inflow will affect the operation of waterjet propulsion pump, which will not only reduce the efficiency of the pump, but also cause unsteady vibration and noise. The uniformity of the inlet flow field of the jet propulsion pump is closely related to the uniformity of the outlet of the inlet duct, The bending characteristics of the inlet duct and the influence of the shaft obviously influent the flow uniformity.

In this paper, using the axial surface velocity non-uniformity coefficient ξ to judge the uniformity of the flow, ξ are defined as follows equation (Eq. 3).

$$\xi = \frac{1}{Q} \int_{dA} \left| u - \overline{U} \right| dA \qquad (3)$$

Where, Q is the volume flow of the section; U is the axial velocity of each element on the section. Fig. 5 shows the locations to analysis outflow of the inlet ducts.

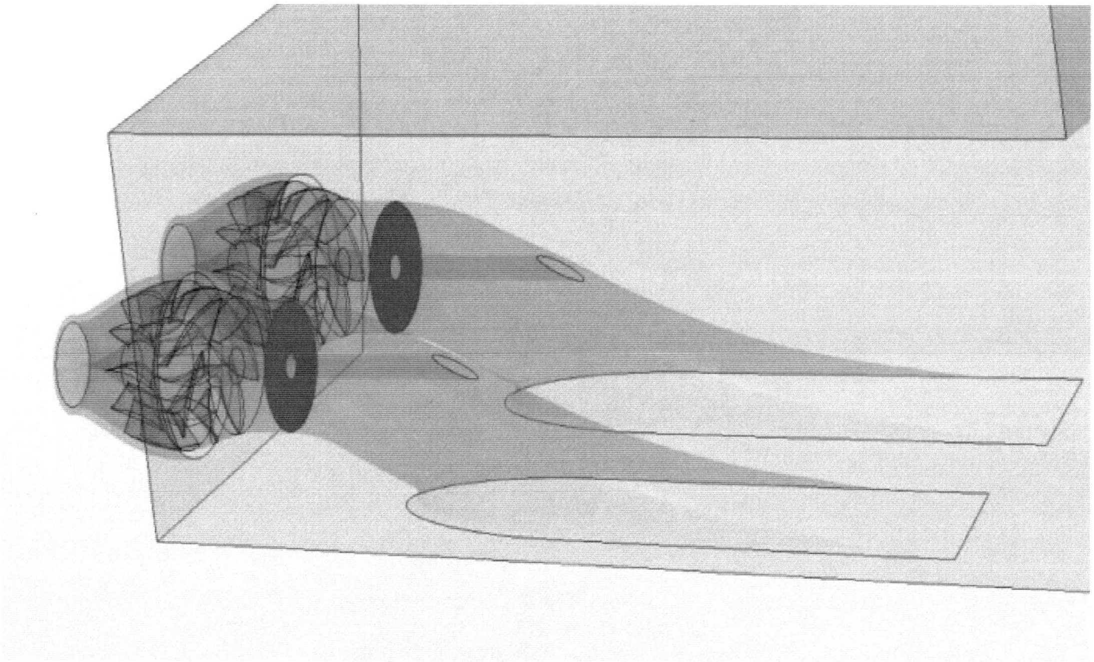

Fig. 5 The locations to analyse outflow of the inlet ducts

In order to reduce the influence of impeller rotation to the outflow of the ducts in the calculation, the section is cut as far as possible from the plane of the pump. In order to analyse the uniformity of the outflow of the inlet duct caused by the deadrise angle of the stern, two kinds of impeller rotation conditions were calculated. Impeller is not uniformity coefficient of entrance section is deduced, the greater the corresponding section of the velocity field is non-uniformity; Factor ξ is smaller, the corresponding section of the velocity field is more uniform.

Table 2. Comparison of non-uniformity coefficients of outlet flow in different spiral downward inlet duct

Rotation direction	Inner Waterjet	Outer Waterjet
clockwise	0.2087	0.2364
counterclockwise	0.2094	0.2374

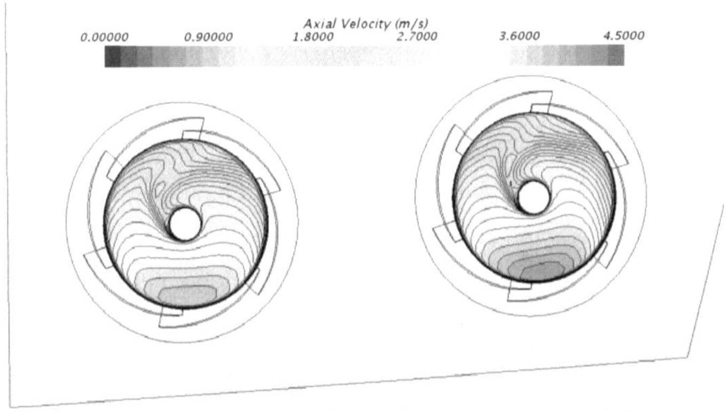

a) clockwise(the left side is the inner impeller)

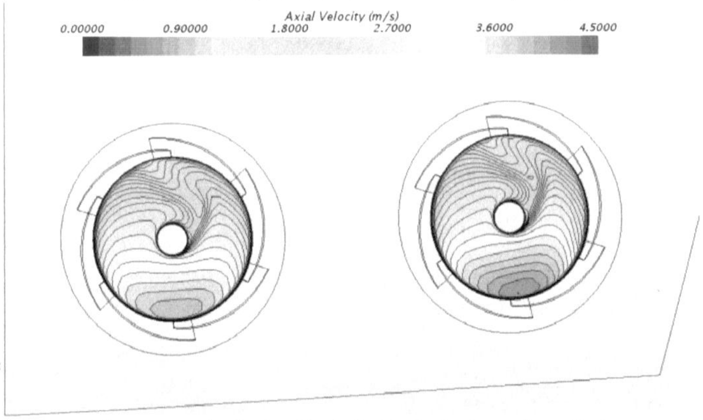

b) counterclockwise(the left side is the inner impeller)

Fig. 6 RANS predicted velocity distributions in impeller inlet under different rotation direction conditions

From different pressure coefficient distributions on impeller surface can be observed (Fig. 7), uneven flows affect the pressure distribution on the surface of the impeller, to the first half of the low flow speed propulsion pump impeller leading edge and there is an obvious low-pressure area, surface area and the lower part of the flow velocity is higher area, promote the pump impeller surface pressure is higher than the upper part. The lowest pressure position on the impeller surface is related to the minimum velocity position of the incoming flow by comparing the pressure cloud diagram of the impeller surface in left-right rotation condition. The size and direction of the outflow velocity are affected by the working conditions. The uneven flow field at the pump inlet causes the pulsation of the impeller's single channel flow. In addition, the load on the impeller blade also shows the characteristics of pulsation due to the influence of non-uniform incoming flow. When the blade turns above the shaft, the load is the largest; when it turns below the shaft, the load is the smallest.

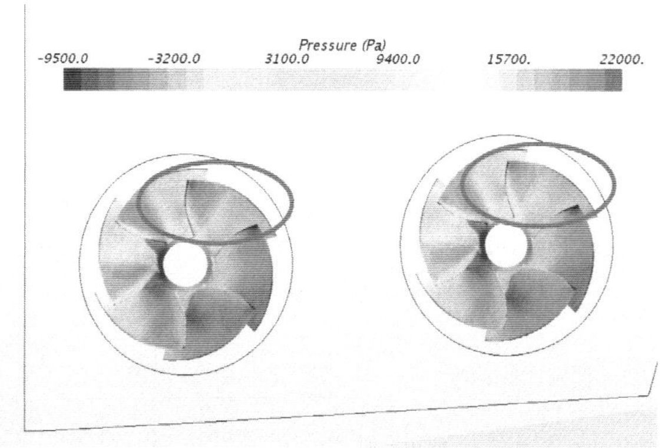

a) clockwise (the left side is the inner impeller)

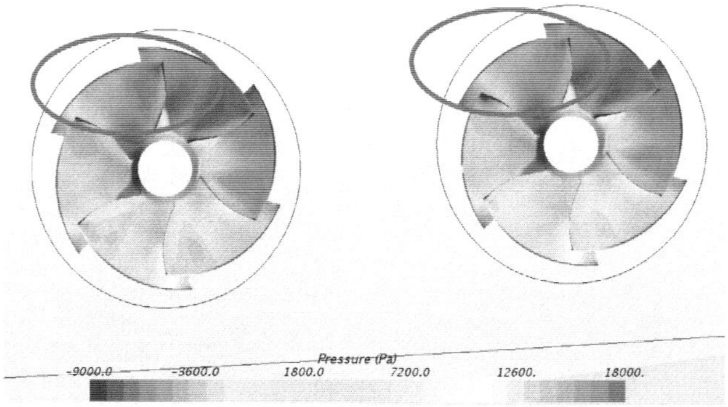

b) counterclockwise (the left side is the inner impeller)

Fig. 7 RANS predicted non-cavitating pressure coefficient distributions under different rotation direction conditions

4. PRESSURE FLUCTUATION RESULTS

Monitoring points

The main purpose of the spectrum analysis of pressure fluctuation is to transform the time domain signal into the frequency domain signal and analyse the frequency components of the signal. Since the pressure fluctuation signal is a stable and random process of each state, it can be used as a sample to represent the set of random processes. At present, Fourier transform, and wavelet transform are widely used in spectrum analysis. In order to ensure the accuracy of the signal and avoid the errors generated in the process of signal acquisition, the signal needs to carry out data pre-processing before spectrum analysis, and then use the method of Fourier transform to expand the Fourier transform of the processed signal sample and extract its main spectrum.

According to the result of spectrum analysis, the performance of the pressure fluctuation in what frequency is stronger can be analysed, and specific analysis of the causes of the main frequency spectrum, the relationship between the head and the measuring point position, which can effectively avoid in the process of actual operation in pressure fluctuation under the condition of operation, increase the safety and reliability of the waterjet propulsion ship.

In order to analyse the pressure fluctuation changes inside the impeller, pressure fluctuation monitoring points were set at 5 sections of the waterjet, which are duct outlet (position A), impeller inlet (position B), impeller tip clearance (position C), impeller outlet (position D) and stator outlet (position E), and 8 monitoring points were arranged clockwise at each section, as shown in the Fig. 8.Eight pressure monitors are placed at the same locations as the pressure transducers, as shown in the Fig. 9. The rotational direction of impeller is clockwise in the following analysis.

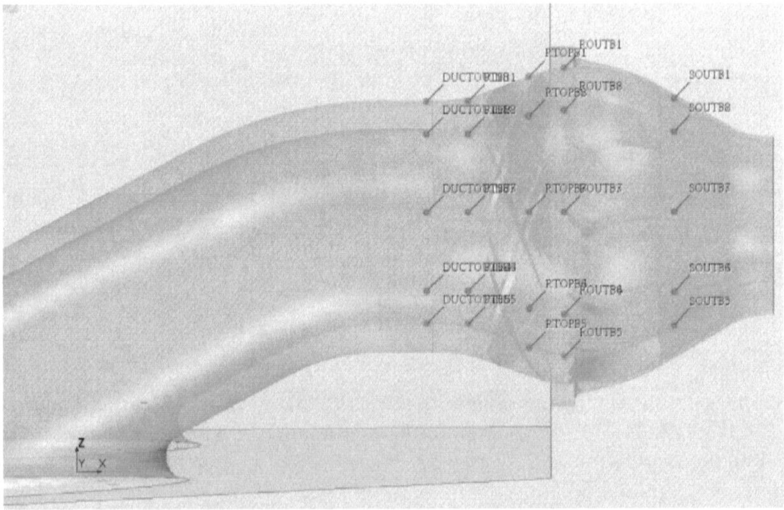

Fig. 8 Pressure fluctuation monitoring points were conducted at 5 sections of the waterjet

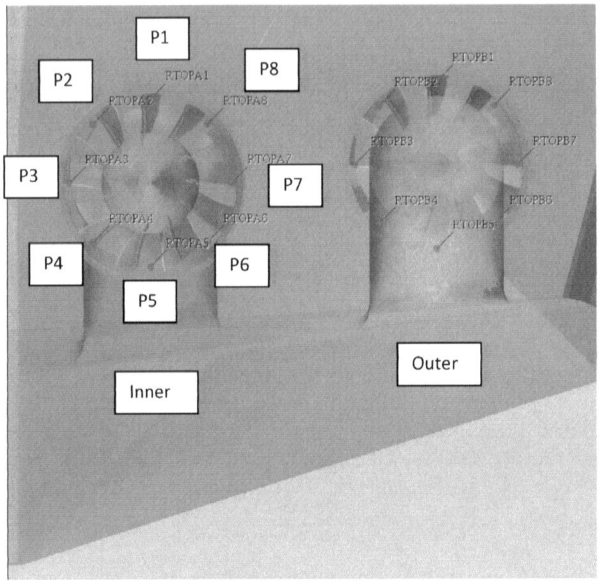

Fig. 9 Eight monitoring points were arranged clockwise at each section

Pressure fluctuation

The time-domain signals of each pressure measurement point obtained by three-dimensional transient calculation are shown in the figure 10. The Fast Fourier Transform (FFT) is used to analyse the pressure fluctuation spectrum of each monitoring point. The Fig.11 shows the pressure coefficient frequency domain diagram of 8 measuring points at the clearance of the top of the impeller. The frequency corresponding to the pressure fluctuation peak is the frequency of the impeller cotyledon and its doubling frequency, and the peak pressure coefficient decreases gradually with the increase of frequency. Pressure coefficients are defined as follows equation (Eq. 4), the P in the equation is the pressure fluctuation of the Fourier transform analysis, D is the impeller diameter, n is the rotation rate.

$$C_p = \frac{P}{\rho \times n^2 \times D^2} \tag{4}$$

The pressure fluctuation of impeller inlet, impeller blade channels, impeller outlet, Guide vane channels, nozzle and tip of impeller is supervised and discussed, and the difference caused by the non-uniformity is also analysed.

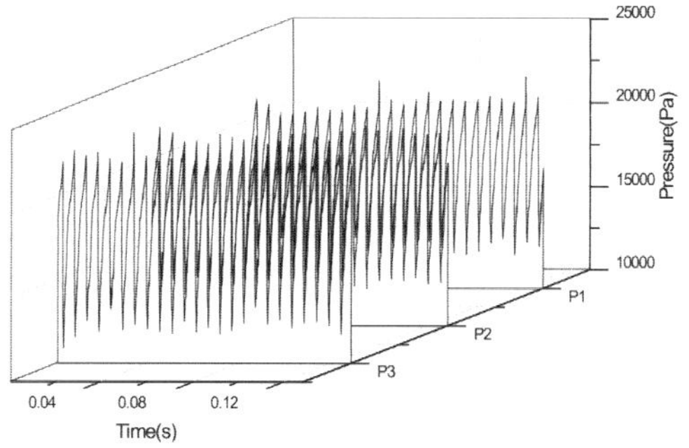

Fig. 10 Time domain signals of pressure measurement points obtained by transient calculation

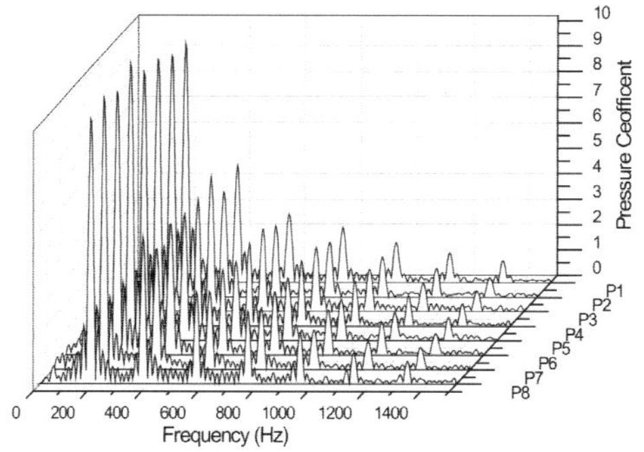

Fig. 11 The frequency domain diagram of the pressure fluctuation coefficient at the tip clearance is obtained by spectrum analysis

As can be seen from the comparison diagram of the maximum pressure fluctuation coefficients of the five sections, the maximum pressure fluctuation value of the internal jet is located at the impeller tip clearance, where the pressure fluctuation amplitude is much higher than that of the other sections due to the influence of multiple factors such as the tip load of the mixed-flow pump and tip clearance vortex. The inlet pressure fluctuation of the impeller is higher than that of the outlet of the impeller. The inlet of the impeller is affected by the interaction between the non-uniform incoming flow and the low-pressure area on the suction surface of the impeller, and the disturbance of the pressure fluctuation is higher than that of the stator. The pressure fluctuation amplitude at the inlet and outlet of the stator is very small, less than 10% relative to the impeller area.

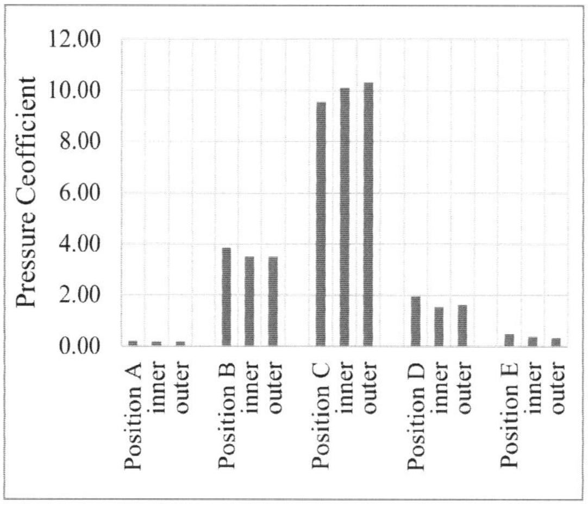

Fig. 12 The comparison of the maximum pressure fluctuation coefficients in the five sections on each position

The inlet area of the waterjet impeller is the outlet of the inlet duct. The disordered inlet flow will affect the normal operation of the waterjet pump, which not only reduces the minimum pressure on the blade surface, but also affects the pressure fluctuation. Since the flow velocity in the top region is lower than that in the bottom region, cavitation is more likely to occur in the top region of the impeller, and the pressure fluctuation here is higher than that in the bottom region. The pressure fluctuation of the inner and outer thrusters is similar.

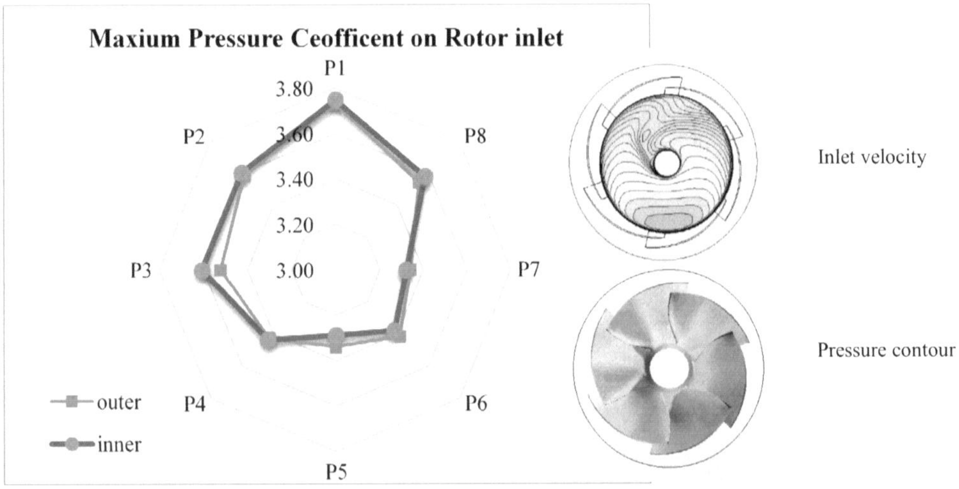

Fig. 13 Maximum pressure fluctuation coefficient on impeller inlet of different points

Compared with the maximum pressure fluctuation coefficient of the 8 measuring points at the tip clearance, the pressure fluctuation in the bottom region of the impeller is higher than that in the top region, which is due to the non-uniformity of the incoming flow, the incoming flow velocity in the bottom region is higher than that in the top region, causing the increasing load on the blades, and the pressure fluctuation increases accordingly. The maximum pressure fluctuation coefficient of the outer thrusters is higher than that of the inner thrusters, the amplitude is about 16%, the axial height of the inner and outer thrusters is slightly different, and the impeller load caused by the interaction of different incoming flow non-uniformity is different.

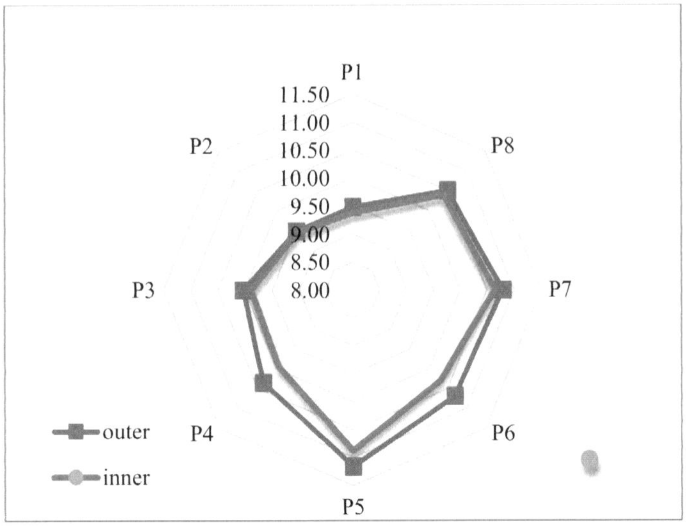

Fig. 14 Maximum pressure fluctuation coefficient on impeller tip of different points

The maximum pressure fluctuation coefficient at impeller outlet is lower than that at impeller inlet and far lower than that at impeller tip clearance. In addition, the pressure fluctuation amplitude of the monitoring point on the right side and the monitoring point on the left side have a relatively large change, which seems to be affected by the rotation direction of the impeller, so it is necessary to make a comparative analysis of the rotation direction change in the future research.

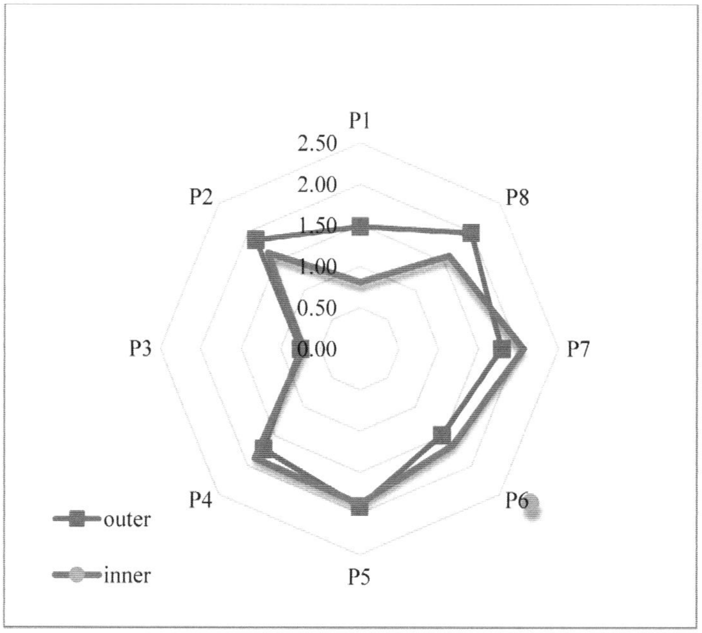

Fig. 15 Maximum pressure fluctuation coefficient on impeller tip of different points

5. CONCLUSIONS

Through the three-dimensional steady and unsteady numerical simulation of the 4 waterjets installed in the aft of a v-shaped ship, the duct outflow characteristics of the internal and external waterjets were obtained, and the circumferential pressure fluctuation characteristics of the shell at the inlet of the impeller, the clearance between the top of the impeller and the outlet of the impeller and the nozzle of the stator were analysed in the paper.

1. The axial flow velocity of the inlet duct of the impeller decreases gradually from bottom to top, with the same trend of the inner and outer thrusters. In addition, the axial velocity of the inlet duct is obviously affected by the rotation of the jet propulsion shaft, especially in the upper part of the shaft where the velocity is lower, and the low-speed region is significantly increased. The uneven inlet flow will affect the pressure distribution on the impeller surface. In the upper part of the impeller with lower inlet velocity, there are obvious low-pressure areas in the leading edge and suction surface of the impeller, while in the bottom of the impeller with higher inlet velocity, the surface pressure of the impeller is higher than that of the upper part.
2. The main frequency of pressure fluctuation in the propulsion pump is the passage frequency of the impeller. Therefore, the natural mode of the main structure should avoid integer multiples of the passage frequency of the blade to improve the operation stability. As can be seen from the comparison diagram of the maximum pressure fluctuation coefficients of the five sections, the maximum pressure fluctuation value of the waterjets is located at the tip clearance, where the pressure fluctuation amplitude is much higher than that of the other sections due to the influence of multiple factors such as the tip load of the mixed-flow pump and tip clearance vortex. Therefore, the shell structure needs strengthening on the impeller region.
3. The pressure fluctuation of the impeller inlet position is higher than that of the outlet of the impeller. The inlet of the impeller is affected by the interaction between the un-uniform inlet flow and the low-pressure area on the blade surface at the inlet of the impeller. The disordered inlet flow field will affect the normal operation of the jet pump, which not only reduces the minimum pressure on the blade surface, but also affects the pressure fluctuation. The pressure fluctuation of the inner and outer thrusters is similar.
4. The circumferential amplitude of pressure fluctuation presents different characteristics at different positions: the highest point of pressure fluctuation amplitude at the inlet of the impeller is located at the top of the minimum velocity on the axial surface, and the highest point of pressure fluctuation amplitude at the clearance between the impeller and the shell is located at the bottom of the maximum velocity on the axial surface.

5. The impeller load difference caused by the interaction of slightly different axial heights of the inner and outer thrusters and the non-uniformity of the incoming flow, the maximum pressure fluctuation coefficient of the outer thrusters is higher than that of the inner thrusters, and the amplitude is about 16%.5. The pressure fluctuation amplitude of the monitoring point on the right side and the monitoring point on the left side both changed significantly, which seemed to be affected by the rotation direction of the impeller, so a comparative analysis of the rotation direction change was needed.

6. REFERENCES

1. Bensow, R. E., & Gustafsson, R. (2017). "Effect of propeller tip clearance on hull pressure pulses". *Proceedings of the 5th International Symposium on Marine Propulsors.*

2. Hirt, C.W., Nichols, B.D., 1981. Volume of fluid (VOF) method for the dynamics of free boundaries. *Journal of Computational Physics.* 39 (1), 201–225

3. Kandasamy, M., Peri, D., Tahara, Y., Wilson, W., Miozzi, M., Georgiev, S. & Stern, F. (2013). "Simulation based design optimization of waterjet propelled Delft catamaran". *International Shipbuilding Progress*, 60(1-4), 277-308.

4. Takai, T., Kandasamy, M., & Stern, F. (2011). "Verification and validation study of URANS simulations for an axial waterjet propelled large high-speed ship". *Journal of marine science and technology*, 16(4), 434-447.

5. Zobeiri, A., Kueny, J. L., Farhat, M., & Avellan, F. (2006). "Pump-turbine impeller-stator interactions in generating mode: pressure fluctuation in distributor channel". In *Proceedings of the 23rd IAHR Symposium on Hydraulic Machinery and Systems, Yokohama, Japan,* 1, 1-10.

NUMERICAL SIMULATION AND EXPERIMENTAL VERIFICATION OF THE NOZZLE FLOW OF A WATERJET MODEL

Z Zhang[1] and **C Yang**, Shanghai Jiao Tong University, China
G Chen, [1]Marine Design & Research Institute of China, China
J Gong, Wuhan University of Technology, Wuhan, Hubei, China

ABSTRACT

The waterjet thrust is calculated usually according to the momentum theorem by obtaining the momentum flux at the inlet and nozzle. Most of the relevant studies focus on the inlet flow, but less on the nozzle flow. The conventional method is to assume that the nozzle flow is uniform, which is different from the actual situation and may affect the accuracy of waterjet thrust prediction. In this paper, A waterjet flow was calculated by RANS method and verified by the experiment. Finally, the differences between the real flow profile and the conventional assumed one are compared, and the influence on the nozzle momentum and the thrust are analyzed. According to the results, the axial momentum flux of the non-uniform flow is greater than that of the uniform one.

1. INTRODUCTION

It is important to calculate or measure the thrust for predicting the waterjet performance. The 21st ITTC in 1996 recommended two methods to measure the thrust of waterjet: momentum flux method and direct thrust measurement method (*Proceedings of the 21st ITTC, 1996*). The direct thrust measurement method is difficult to realize in the experiment. Most research institutions used momentum flux method to obtain the thrust of waterjet. The so-called momentum flux method regards the fluid flowing through the waterjet unit as a control volume. The force on the control volume, that is the thrust of the waterjet unit, is calculated according to the momentum theorem by obtaining the momentum flux at the inlet and nozzle. The conventional method is to assume that the nozzle flow uniform (*Proceedings of the 23st ITTC, 2002*), and then calculate the nozzle momentum flux when the nozzle diameter is known. However, the jet flow is not uniform actually due to the influence of the nozzle profile, which will cause the difference between the calculated momentum and the actual one, and further affect the prediction accuracy of waterjet thrust. Scherer, O. (2001) has used Laser Doppler velocimetry to determine the momentum flux for calculating the thrust more accurately. Park, W.G. et al (2005) computed the jet velocities just behind the nozzle and compared them with the experiment results. The 24th ITTC provided the jet velocity survey data by six research institutes. The jet survey was measured with Laser Doppler, Pitot-static and 5-hole probes. (*Proceedings of the 24th ITTC, 2005*). Jung, K.H. et al (2006) measured velocity vector fields at the nozzle exit using stereoscopic PIV measurements.

In this paper, the internal and external flow of a waterjet model was numerically simulated based on RANS equation. The free surface was considered. The experiment about measuring of the nozzle flow of the waterjet model was carried out in the Science and Technology on Water Jet Propulsion Laboratory, Marine Design & Research Institute of China. The pitot tube probe was used to measure the axial velocity at each radius within the nozzle section. The numerical results were verified by comparing with the experimental results. Finally, the differences between the actual nozzle flow and the assumed uniform nozzle flow were compared, and the influences on the nozzle momentum flux and the thrust were analyzed.

According to the results, the numerical results are consistent with the experimental results, and the actual nozzle momentum flux is greater than that calculated by the uniform nozzle flow assumption, which causes the actual thrust is greater than the uniform one. So, the waterjet thrust predicted by the uniform nozzle flow assumption is more conservative.

2. RESEARCH OBJECT

In this paper, a mixed-flow pump model was selected as the research object to study the nozzle flow characteristics. The mixed-flow pump model is composed of inlet duct, shaft, impeller, guide vane and nozzle. The 3D model is illustrated in Fig. 1, and the characteristic diameter (inlet duct diameter before impeller) is 0.2m.

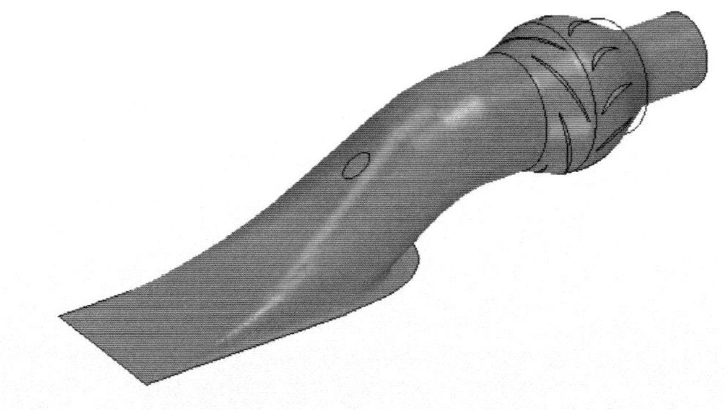

Fig. 1. The mixed-flow waterjet pump 3D model

A simple hull model with a flat bottom was used in this research, because the hull shape has little effect on the nozzle flow. The ship model is 3.2m long, 1.0m wide, 0.6m high and 0.2m draft. The 3D model of waterjet ship model is shown in Fig. 2.

Fig. 2. The waterjet ship 3D model

3. NUMERICAL SIMULATION

Calculation Domain
Star-CCM+ was used for the numerical simulation of the flow. Firstly, the calculation domain including hull, waterjet pump and the flow around was established. The geometry size of the domain is shown in Fig. 3.

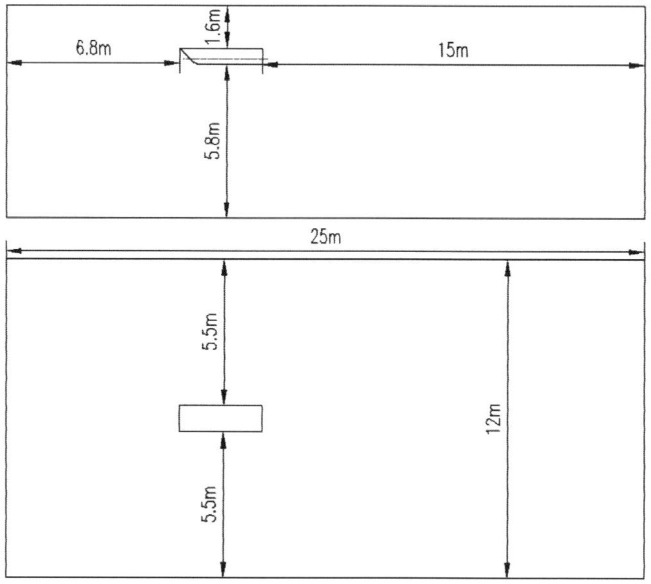

Fig. 3. The geometry size of the calculation domain

The impeller and the flow around were set as the rotating domain and placed into the rotating reference frame. The other areas including the inlet duct, guide vane, nozzle and hull were set as the stationary domain, as shown in Fig. 4.

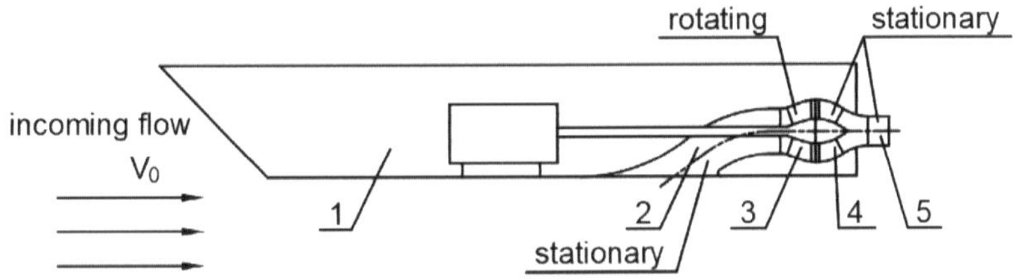

Fig. 4. Calculation domain settings
1 hull, 2 inlet duct, 3 impeller, 4 guide vane, 5 nozzle.

Numerical Setup

RANS method and the SST k-ω turbulence model were applied for this simulation. The wall function method was applied to deal with the near wall flow, and the free surface was captured by gas-liquid two-phase flow model and VOF plane wave model.

The flow in front of the impeller is non-uniform, so the calculation is an unsteady problem. Considering the influence of time step on the calculation results, the time step was set as 1/300s, 1/600s and 1/1200s respectively for the time uncertainty analysis below.

The velocity inlet, pressure outlet and fixed wall were applied for the boundary conditions. The velocity inlet boundary was set at the inlet, both sides and top of the calculation domain. The pressure outlet boundary was set at the outlet. The fixed wall boundary is set on the other walls.

The calculation cases are listed in Table 1.

Table 1. The calculation cases

Case	Incoming flow velocity (m/s)	Rotating speed (rpm)
1	2	400
2	2	500
3	2	600
4	2	640

Mesh

Considering the influence of grid scale on the calculation results, it is necessary to analyze the grid uncertainty. In this calculation, trim grid was applied in each domain, which belongs to unstructured grid. A total of four sets of grids with different scales were generated for the grid uncertainty analysis.

The basic size of the grid was set to 0.7m, 1m, 1.4m and 2m respectively. Blocks are added to the free surface, around the hull, inside the water jet propulsion pump and near the inlet and outlet to encrypt the grid. The dimensions of the grids in different areas are given in the form of percentage of the basic dimensions to ensure that the ratio of the overall dimensions of the four sets of grids is about. At the same time, boundary layer grids are added to the wall areas such as pipe wall, blade surface and hull surface of water jet propulsion pump.

4 meshes are shown in Fig. 5. It is the thinnest to the coarsest from top to bottom.

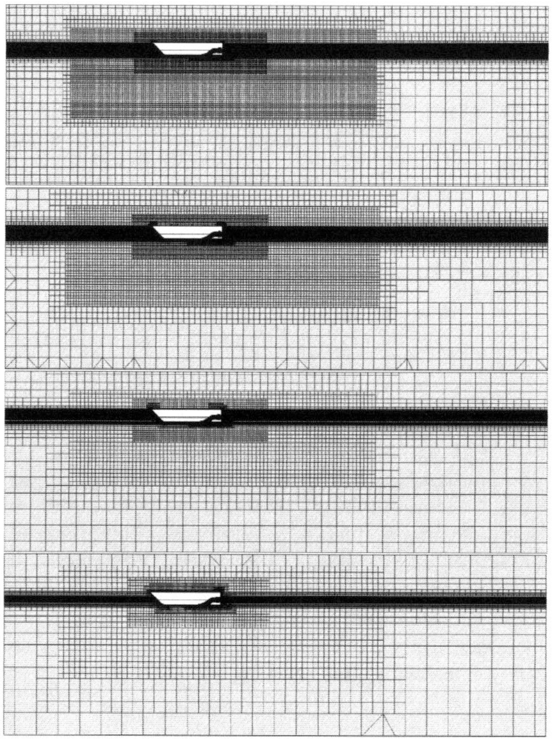

Fig. 5. 4 meshes with different scales

The mesh quantity is listed in Table 2.

Table 2. The quantities of the 4 meshes

No.	Mesh scale	Mesh quantity (Unit: million)		
		Rotating domain	Stationary domain	total
1	thinnest	1.092	13.08	14.172
2	thin	0.646	5.653	6.299
3	coarse	0.328	2.969	3.297
4	coarsest	0.0181	1.489	1.670

Uncertainty Analysis

The uncertainty was analyzed according to the ITTC – Recommended Procedures and Guidelines, "Uncertainty Analysis in CFD Verification and Validation, Methodology and Procedures" (2017). The expression of numerical uncertainty is as Eq.1.

$$U_{SN}^2 = U_I^2 + U_G^2 + U_T^2 + U_P^2 \quad (1)$$

Where, U_I is the iteration uncertainty;
U_G is the grid uncertainty;
U_T is the time uncertainty;
U_p is the uncertainty caused by other factors, which is not considered and ignored in this calculation.
Case 2 in Table 1, where incoming flow speed is 2m/s and rotating speed is 500rpm, was selected to calculate the thrust of the waterjet pump at first. And then the numerical uncertainty of the thrust was analyzed.

Iteration uncertainty U_I

The expression of iteration uncertainty is as Eq. 2.

$$U_I = \frac{1}{2}(S_{\max} - S_{\min}) \tag{2}$$

Where, S_{max} is the maximum of the results in the final iteration process.
S_{min} is the minimum of the results in the final iteration process.
According to the thrust results, S_{max}=157.45N， S_{min}=153.49N.
And then, U_I=1.98=1.27%S_T

Grid uncertainty U_G

U_G is the grid uncertainty. Because the unstructured mesh was used in this calculation, the least square method (LSR) and grid exponential convergence method (GCI) was used to solve the uncertainty.
The solution of uncertainty can be expressed by Eq. 3.

$$S_i = S_0 + \alpha h_i^p \tag{3}$$

Where, S_i is the calculation result of each mesh, which is the waterjet thrust in this case.
S_0 is the theoretical solution assuming that the grid scale is infinitesimal.
h_i is the grid scale. Because the scales ratio of unstructured grids is difficult to be strictly taken as $\sqrt{2}$, h_i was taken as the reciprocal of the third root of the total grid quantity in this calculation.
p is a parameter indicating the convergence of the results. P>0 indicates monotonic convergence and P<0 indicates divergence.
The unknown variables of S_0, α, and p were solved by fitting the results of the four meshes according to the least square method. And then the grid uncertainty was calculated by Eqs. 4~5.

$$U = F_s|\delta_{RE}| \tag{4}$$

$$\delta_{RE} = S_i - S_0 = \alpha h_i^p \tag{5}$$

Where, F_s is the safety factor. When $0.5 \le P \le 2$, F_s is taken as 1.25. When p is equal to other values, F_s is taken as 3.
The thrusts calculated by the 4 meshes are shown in Table 3.

Table 3. The thrusts calculated by the 4 meshes

No.	Mesh scale	Mesh total quantity (Unit: million)	Thrust (N)
1	thinnest	14.172	152.4
2	thin	6.299	155.7
3	coarse	3.297	155.6
4	coarsest	1.670	159.8

According to the least square method, the solutions of Eq. 3 are as follows:
S_0=149.2， p=1.485， α=130.
Because $0.5 \le p \le 2$， F_s=1.25.
And then,

$$U_G = F_s|\delta_{RE}| = F_s|\alpha h_i^p| = 4.29\% S_T \tag{6}$$

Time uncertainty U_T

U_T is the time uncertainty. Because the calculation is unsteady, the convergence of time steps should be considered. The convergence factor formula is as Eq. 7.

$$R_T = \frac{\varepsilon_{21}}{\varepsilon_{32}} = \frac{S_2 - S_1}{S_3 - S_2} \tag{7}$$

Where, S_1, S_2 and S_3 are the calculated thrusts when the time steps are 1/1200s, 1/600s and 1/300s respectively.
R_T is the convergence factor. When $0 < R_T < 1$, the results are monotonic convergence. When $R_T < 0$, the results are oscillatory convergence. When $R_T > 1$, the results are divergence.
Accuracy order p_T and error estimation δ_{RET} are calculated by Eqs. 8~9.

$$p_T = \frac{\ln(\varepsilon_{21}/\varepsilon_{32})}{\ln(r_T)} \tag{8}$$

$$\delta_{RET} = \frac{\varepsilon_{21}}{r_T^{p_T} - 1} \quad (9)$$

Where, r_T is the ratio of time steps, which is equal to 2 in this calculation.
$\varepsilon_{21} = S_2 - S_1$, $\varepsilon_{32} = S_3 - S_2$.
The correction factor C_T is calculated by Eq. 10.

$$C_T = \frac{r_T^{p_T} - 1}{r_T^{p_{Test}} - 1} \quad (10)$$

Where p_{Test} is the format accuracy, which is equal to 2 in this paper.
The time steps are 1/1200s, 1/600s and 1/300s respectively. The calculation results are listed in the Table. 4.

Table. 4 The thrusts calculated by different time steps

No.	Time step (s)	Rotation angle per time step (°)	Mesh quauntity (million)	Thrust （N）
1	1/1200	2.5	6.299	155.7
2	1/600	5	6.299	156.19
3	1/300	10	6.299	156.77

The time uncertainty calculation results are listed in Table. 5.

Table. 5 The time uncertainty calculation results

ε_{21}	ε_{32}	r_T	R_T	p_T	δ_{RET}	C_T	$U_T(\%S_T)$
0.49	0.58	2	0.845	0.243	2.668	0.06	0.0161

Because $0 < R_T < 1$, the thrusts are monotonic convergence.
Other uncertainty U_p
The uncertainty caused by other factors U_p is usually not considered. In this paper, it is equal to 0.
In the end, the numerical uncertainty can be calculated as Eq. 11.

$$U_{SN} = \sqrt{U_I^2 + U_G^2 + U_T^2} = 4.76\% S_T \quad (11)$$

It can be considered that the numerical calculation results of thrust are satisfactory.

Numerical Results

In the end, No. 3 mesh was used for the nozzle flow calculation in all of the cases, and the time step was set as 1/600s. The axial velocity profiles on the section 50mm upstream of the nozzle in different cases are illustrated in Fig. 6.

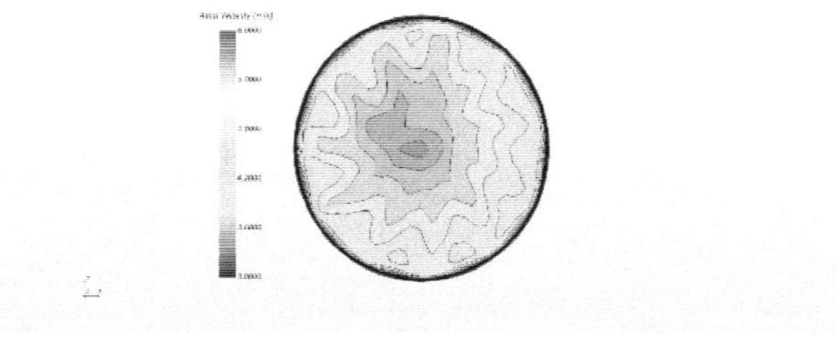

a. 400rpm

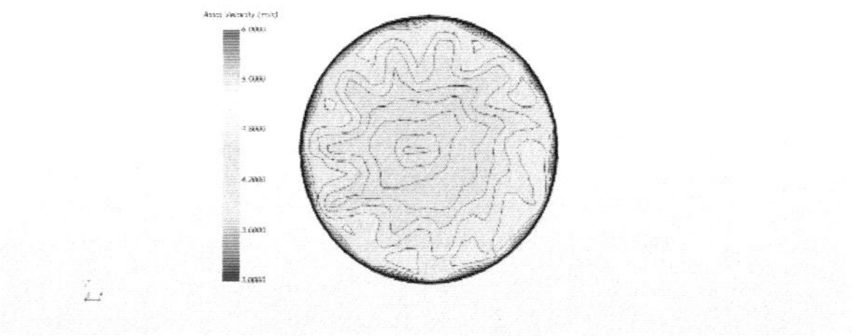

b. 500rpm

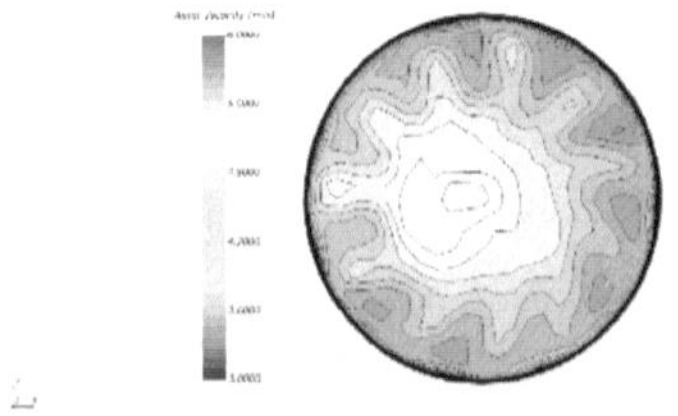

c. 600rpm

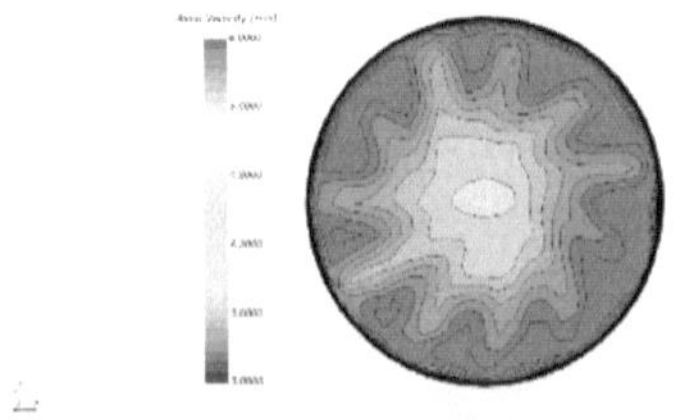

d. 640rpm

Fig. 6. Axial velocity profiles on the nozzle section in different cases

4. EXPERIMENT VARIFICATION

Experiment Setup

The waterjet nozzle flow test was carried out in the circular channel with free surface in the Science and Technology on Water Jet Propulsion Laboratory, Marine Design & Research Institute of China. According to the research object, the 1:1 test model is processed. The test model is composed of the hull, motor, torque meter, waterjet pump and pitot tubes, as shown in the Fig. 7. The test model is shown in Fig. 8.

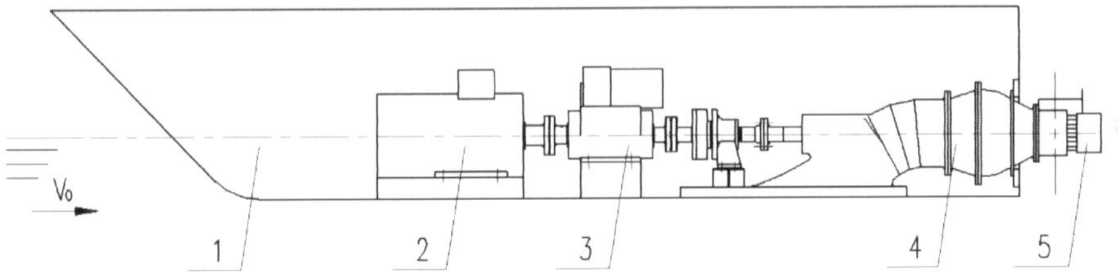

Fig. 7. Test model composition
1. Hull, 2. Motor, 3. Torque meter, 4. Waterjet pump, 5. Pitot tubes.

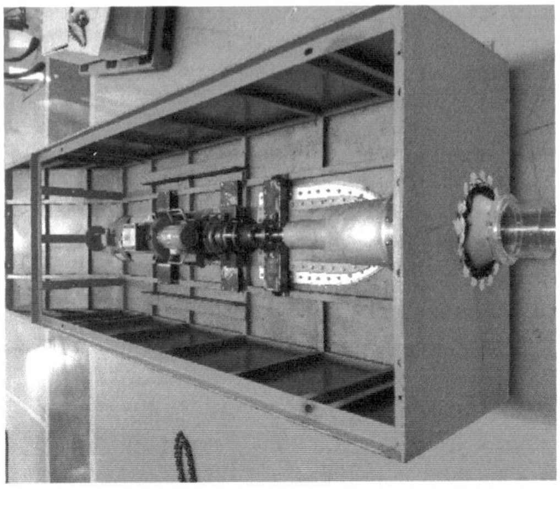

Fig. 8. The photo of the test model

The pitot tubes consist of 9 probes and arrange in a straight line. The probes extend into the nozzle to measure the axial velocities of 9 points on a diameter in the nozzle cross-section. By rotating the pitot tubes along the central axis, the axial velocities on different angular diameters in the section can be measured. The measurement positions are shown in Fig. 9 (unit: mm):

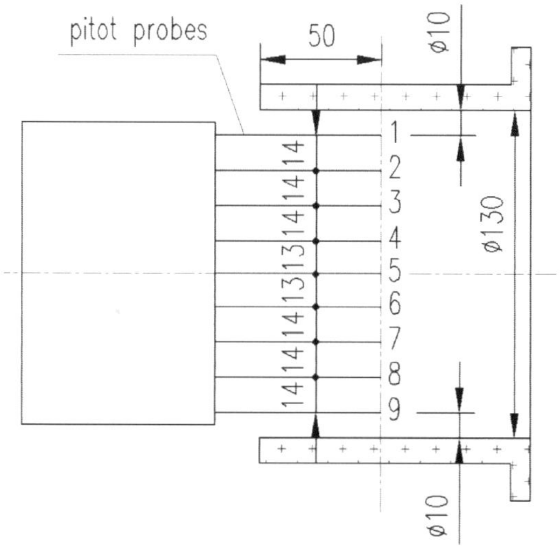

Fig. 9. The measurement positions of the pitot probes

Experiment method

During the experiment, the waterjet pump operates at a given rotating speed at a certain incoming flow velocity. The test cases are the same with Table 1.

In each case, the axial velocities in the four directions of 0°, 30°, 60° and 90° were measured and recorded respectively. Figure 10 shows the tail flow phenomenon of one case during the experiment.

Fig. 10. The tail flow in the experiment

5. RESULTS DISCUSSION

Results Comparison

The numerical calculated axial velocities in the four directions of 0°, 30°, 60° and 90°, consistent with the test positions, were read and drawn in the same diagram of curves with the test results, as shown in the Figs.11~14.

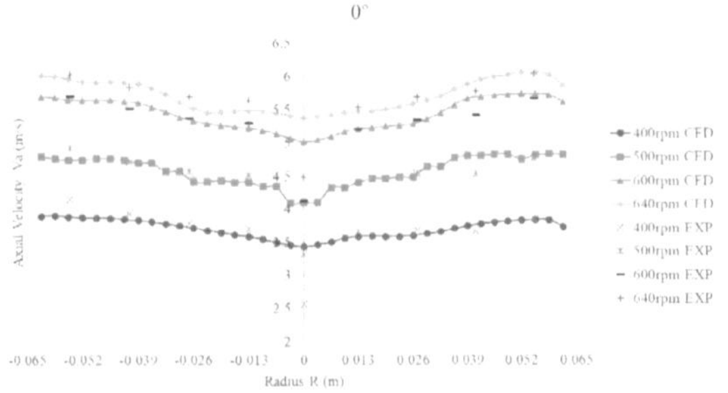

Fig. 11. The axial velocity curves in 0° direction

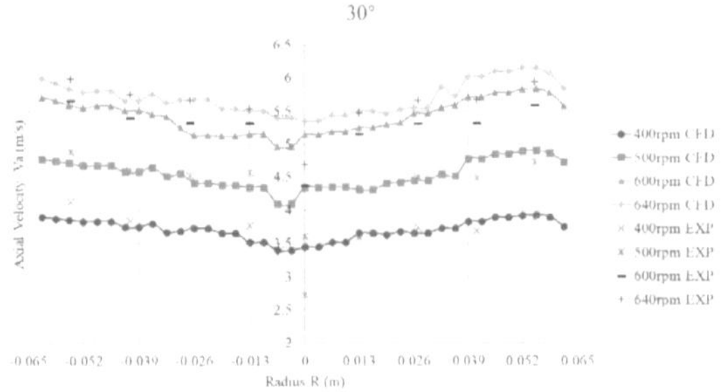

Fig. 12. The axial velocity curves in 30° direction

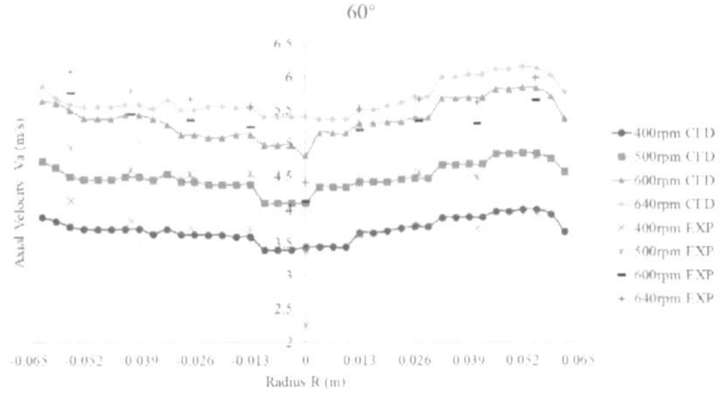

Fig. 13. The axial velocity curves in 60° direction

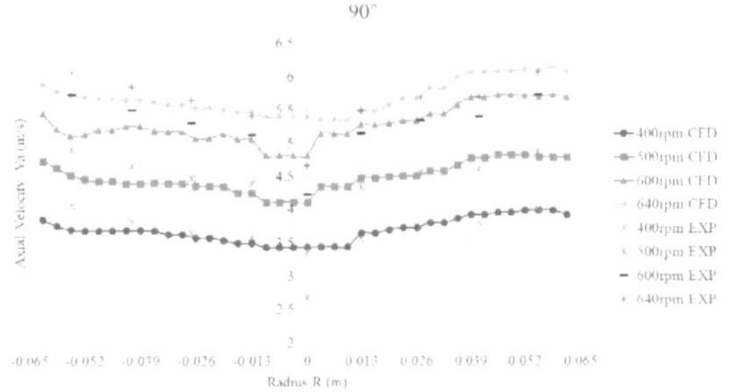

Fig. 14. The axial velocity curves in 90° direction

It can be found that the calculation results are in good agreement with the test results in most areas of the section, which proves that the calculation results are reliable. However, there is a large deviation of the velocity near the center of the nozzle section. It is probably because that the local grid scale was too large to simulate the flow confluence vortices near the center line better.

Comparison with Uniform Nozzle Flow

The waterjet thrust is usually calculated by subtracting the inlet momentum flux from the nozzle momentum flux, as shown in Eq. 12.

$$T = M_j - M_i \qquad (12)$$

Where, T is the waterjet thrust;
M_j is the nozzle momentum flux;
M_i is the inlet momentum flux.

The influence of inlet velocity distribution on inlet momentum flux is not considered in this paper. Assuming that the waterjet pump incoming flow is uniform and the inlet boundary layer influence is ignored, the inlet momentum flux of the waterjet pump can be calculated by Eq. 13.

$$M_i = \rho Q V_0 \qquad (13)$$

Where, ρ is the liquid density;
Q is the rate of flow;
V_0 is the incoming flow velocity.

The nozzle momentum flux is generally calculated based on the assumption of uniform flow in the conventional engineering calculation, as shown in Eq. 14.

$$M_{j1} = \rho Q \overline{V_j} = \rho Q^2 / A_j \qquad (14)$$

Where, M_{j1} is the uniform nozzle momentum flux;
$\overline{V_j}$ is the average nozzle axial velocity based on the uniform flow assumption, $\overline{V_j} = Q/A_j$;
A_j is the nozzle area.

And then the waterjet thrust of uniform nozzle flow assumption, T_1, can be expressed by Eq. 15.

$$T_1 = M_{j1} - M_i \qquad (15)$$

It is obvious that the nozzle flow is non-uniform according to the calculation and test results in this paper. When the velocity distribution of the whole nozzle section is known, the actual nozzle momentum can be calculated by the area integral of the velocity square on the section, as shown in Eq. 16.

$$M_{j2} = \rho \int_{A_j} V_j^2 ds \qquad (16)$$

Where, M_{j2} is the actual nozzle momentum flux;
Vj is the actual axial velocity at each position of the nozzle section, obtained by CFD or experiment;

And then the actual waterjet thrust, T_2, can be calculated by Eq. 17.

$$T_2 = M_{j2} - M_i \qquad (17)$$

According to the above contents, the nozzle momentum flux and thrust of each case were calculated and compared with the results of uniform nozzle flow assumption, as shown in Table 6.

Table. 6 Comparison between the actual results and the uniform ones

Variable	symbol	results			
Incoming velocity	V_0(m/s)	2	2	2	2
Rotating speed	n(rpm)	400	500	600	640
Flow rate	Q(m^3/s)	0.04929	0.06041	0.07147	0.07606
Nozzle diameter	D_j(m)	0.13	0.13	0.13	0.13
Nozzle area	A_j(m^2)	0.01327	0.01327	0.01327	0.01327
Uniform nozzle momentum flux	M_{j1}(N)	183.02	274.97	384.81	435.9
Actual nozzle momentum flux	M_{j2}(N)	184.12	276.52	386.93	438.26
Deviation of nozzle momentum flux	σ_1	0.60%	0.56%	0.55%	0.54%
Inlet momentum flux	M_i(N)	98.57	120.83	142.94	152.13
Uniform thrust	T_1(N)	84.44	154.15	241.88	283.77
Actual Thrust	T_2(N)	85.55	155.69	244.00	286.13
Deviation of thrust	σ_2	1.30%	1.00%	0.88%	0.83%

According to the results in the Table. 6, it can be found that the actual nozzle momentum flux is about 0.5% ~ 0.6% higher than the uniform nozzle momentum flux because of the nozzle flow non-uniformity at the same flow rate.

The nozzle momentum flux deviation decreases with the flow rate increasing, which further causes the deviation of waterjet thrust. When the same inlet momentum flux is subtracted, the thrust deviation is magnified to about 0.8% ~ 1.3%.

6. CONCLUSIONS

In this paper, the nozzle flow of a waterjet ship model was numerically simulated and experimentally verified. The actual calculation results were compared with those of the uniform nozzle flow assumption. In the end, the following conclusions were obtained:

The actual nozzle flow is non-uniform. When the flow rate is equal, the non-uniformity causes the actual nozzle momentum flux to be slightly larger than that of uniform flow.

The actual nozzle momentum flux is larger than the uniform nozzle momentum flux, which further leads to greater actual thrust. Therefore, there is an error between the predicted thrust and the actual thrust under the traditional uniform nozzle flow assumption in engineering. This error is conservative and small, so it is acceptable in engineering.

7. ACKNOWLEDGEMENTS

We would like to thank Wang Zonglong and Zhu Jianshen for helping me complete the measurement experiment of nozzle flow.

8. REFERENCES

1. Jung, KH, Kim, KC, Yoon, SY, Kwon, SH, Chun, HH, Kim, MC (2006), "Investigation of turbulent flows in a waterjet intake duct using stereoscopic PIV measurements", *J Mar Sci Technol*, 11:270–278.

2. Park, WG, Jang, JH, Chun, HH, Kim, MC (2005), "Numerical flow and performance analysis of waterjet propulsion system", *Ocean Engineering*, 32(14–15), 1740-1761.

3. Scherer, O, Mutnick, I, and Lanni, F (2001), "Procedure for conducting a towing tank test of a waterjet propelled craft using Laser Doppler velocimetry to determine the momentum and energy flux", 26th ATTC, Webb Institute.

4. The Specialist Committee on Validation of Waterjet Test Procedures (2002). *Proceedings of the 23rd ITTC*, Volume II, 393.The Specialist Committee on Validation of Waterjet Test Procedures (2005). *Proceedings of the 24th ITTC*, Volume II, 495.

5. "Uncertainty Analysis in CFD Verification and Validation, Methodology and Procedures" (2017). *ITTC–Recommended Procedures and Guidelines*, 7.5-03-01-01.

6. Waterjets Group (1996). *Proceedings of the 21st ITTC*, 189-198.

THE ROYAL INSTITUTION
OF NAVAL ARCHITECTS

CONFERENCE PROCEEDINGS

THE ROYAL INSTITUTION OF NAVAL ARCHITECTS

8-9 Northumberland Street, London WC2N 5DA

Tel: +44 (0)207 235 4622 Fax: +44 (0)207 259 5912

conference@rina.org.uk

www.rina.org.uk